36
Ladybugs of Alberta
Finding the Spots and Connecting the Dots

JOHN ACORN

Ladybugs
of Alberta

Finding the Spots and Connecting the Dots

The University of Alberta Press

Published by

The University of Alberta Press
Ring House 2
Edmonton, Alberta, Canada   T6G 2E1

Copyright © John Acorn 2007
Printed and bound in Canada by Houghton Boston Printers, Saskatoon, Saskatchewan.
First edition, first printing, 2007
All rights reserved

LIBRARY AND ARCHIVES CANADA
CATALOGUING IN PUBLICATION

Acorn, John, 1958–
    Ladybugs of Alberta : finding the spots and connecting the dots / John Acorn.

(Alberta insects series)
Includes bibliographical references.
ISBN 978-0-88864-381-0

    1. Ladybugs—Alberta. I. Title. II. Series.

QL596.C65A26 2007   595.76'9097123
C2007-901054-7

No part of this publication may be produced, stored in a retrieval system, or transmitted in any forms or by any means, electronic, mechanical, photocopying, recording, or otherwise, without the prior written consent of the copyright owner or a licence from The Canadian Copyright Licensing Agency (Access Copyright). For an Access Copyright license, visit www.accesscopyright.ca or call toll free: 1-800-893-5777.

The University of Alberta Press gratefully acknowledges the support received for its publishing program from The Canada Council for the Arts. The University of Alberta Press also gratefully acknowledges the financial support of the Government of Canada through the Book Publishing Industry Development Program (BPIDP) and from the Alberta Foundation for the Arts for our publishing activities.

Photo credits: All photos by John Acorn, except for those provided courtesy of Joseph Belicek (p. 31 bottom right), John and Bert Carr (p. 31 bottom left), Dwayne Dorland (p. 32 top right), Steve Marshall (p. 60, 64, 96, 146), Diane Olsen (p. 10), Heather Proctor (p. 150), Mira Snyder (p. 32, bottom), Dena Stockburger (p. 158), and the University of Alberta Strickland Library (p. 28).

To Matt Chew—an insightful, intelligent, and courageous fellow.

A typical two-spot ladybug, one of the most familiar species in North America as well as in Europe.

Contents

Acknowledgements IX
Preface XI
Gallery of the Lesser Ladybugs of Alberta XVII
Gallery of the Larger Ladybugs of Alberta XXIII

1 What is a Ladybug? 1
2 The Life of a Ladybug 13
3 Ladybug Study in Alberta 29
4 Introduced Ladybugs and Conservation 39
5 The Lesser Ladybugs of Alberta 57
6 The Larger Ladybugs of Alberta 99

Appendix A: Checklist 155
Appendix B: Glossary 159
Appendix C: Helpful Sources for Ladybug Study 163
References 165

A melanic or dark-morph two-spot ladybug, quite unlike the typical red and black members of the species.

Acknowledgements

In developing my understanding of Alberta ladybugs, I owe a debt of thanks to a great many people. First, there are my students Patsy Drummond, Mira Snyder, and Wendy Wheatley (maiden name Harrison). Then, there are those who provided records, specimens, or photographs: Robert Bercha, Vanessa Block, John and Bert Carr, Tyler Cobb, Dwayne Dorland, Gerry Hilchie, Steve Marshall, Sarah McPike, Diane Olsen, Freeman Olson, Ted Pike, Rob Roughley, Shelly Ryan, Ken Sanderson, and Cindy Verbeek. Others made their institutional collections available for study, and in this regard I appreciate the help of Felix Sperling and Danny Shpeley (E.H. Strickland Entomology Museum, University of Alberta), Terry Thormin (Royal Alberta Museum), and Greg Pohl (Northern Forestry Centre, Canadian Forest Service). For help with the derivations of scientific names, I thank Selina Stewart. For other discussions, advice, and various offers of help, I would like to acknowledge Karolis Bagdonas, George Ball, Karen Brown, Matt Chew, Chris Fisher, Cedric Gillott, Josh Jacobs, Brian Leishman, David Maddison, Milan and Miles Makale, Dave McCorquodale, Lisa Mervyn, Marie-Pierre Mignault, Edward Mondor, John Obricki, Carroll Perkins, Heather Proctor, Katy Prudic, John Spence, Bill Turnock, Natalia Vandenberg, Lindsay Wickson, and Ian Wise. At the University of Alberta Press, I was delighted to work once more with Alan Brownoff, Linda Cameron, Cathy Crooks, Wendy Johnson, Chau Lam, Michael Luski, Peter Midgley and Yoko Sekiya. Finally, I thank Mike Majerus for the numerous insights he passed on to me, and for continual support and inspiration, I thank my wife Dena and our two boys Benjamin and Jesse.

Top: A lovely male taiga bluet damselfly.
Bottom: An amorous pair of the fletcheri race of the beautiful tiger beetle.

Preface

When I began work on the Alberta Insect Series, it seemed to me that it should be possible to complete one book each year over a five-year period (treating tiger beetles, damselflies, ladybugs, the bigger moths, and dragonflies). That was 7 years ago, and this is only the third book in the series—proof that things become both more complex and more interesting as you start to work on them.

The first book in the series, *Tiger Beetles of Alberta: Killers on the Clay, Stalkers on the Sand* is one of a handful of tiger beetle books that have appeared in recent years. I was glad that it was well received here, that it has found an audience well beyond the boundaries of Alberta, and that the majority of the names I proposed were endorsed by the readers of the journal *Cicindela* for "official" use in more recent field guides, such as the fine book on the Canadian and US tiger beetles by David Pearson, Barry Knisely, and Charles Kazilek. My book also inspired research with students Cindy Sheppard and Randy Dzenkiw on the Alberta populations of the beautiful tiger beetle, which appear to constitute a distinct geographic race. Thus, they deserve their own subspecies name, and we are now championing the resurrection of the name *Cicindela formosa fletcheri* Criddle for these beetles. Make a note in the margin of your tiger beetle book and watch for a scientific paper sometime soon.

Damselflies of Alberta: Flying Neon Toothpicks in the Grass was also one of a number of damselfly guides to appear in recent years, and I was glad to provide an Alberta perspective on these insects. The reason it came second in the series is simple: I had all the photographs I needed for this group of insects, which was not yet the case for the other groups I am planning to feature. The most common feedback I have received on the damselfly book is the question: "When are you going to do the dragonflies, John?" The answer, of course, is "When I get around to it—these books take time!" In the meantime, there are good dragonfly guides available for North America in general and more in the works.

I am delighted to present to you a book on our province's ladybugs. Unlike the first two books in the Alberta Insect Series, the book you now hold is truly alone in its field. Amazingly, there are no other popular guides to the ladybugs of any part of North America! There are, of course, technical works on the subject, but the most important of these (Robert Gordon's monographic papers published in 1976 and 1985) are almost impossible to find unless you have access to a good university library, and difficult to read unless you are a trained entomologist. Another excellent but quite technical reference is Nat Vandenberg's ladybug chapter in the two-volume series "American Beetles." Britain has popular guides to the "ladybirds," but so far we have not been so lucky.

Another difference between this book and the two that preceded it is the number of species covered. There are 19 species of tiger beetles in Alberta, 22 species of damselflies, but a whopping 75 species of ladybugs. This may come as a surprise to most readers, and if so, I hope it is a happy one. To be honest, I'm often surprised by local ladybug diversity myself, and I deeply hope that readers of this book will find themselves inspired to get out and see as many species as possible, rather than being overwhelmed by the number of possibilities when it comes to making an identification.

The underlying theme of my tiger beetle book involves life in isolation, and the dynamics of patchy, highly eroded habitats—places like sand dunes, salt flats, and beaches—kept in existence by geological forces but continually under threat from colonization by pesky plants. Likewise, the damselfly book had a geographic thread running through it, about the appearance of new habitats, such as those created by water management on the prairies, and the warm-water outflows from power plants. The theme for this ladybug book is different—it has to do with the effects of introduced species on a native fauna. Most naturalists are aware that the seven-spot ladybug (*Coccinella septempunctata*) is a newcomer to Alberta, and that other European and Asian species are established elsewhere in North America, and might possibly find their way here as well. In fact, I have had a number of people ask why I would bother with a ladybug book, since "all I see now is that damned seven-spot."

This, to me, is a fascinating story. I too was deeply troubled by the arrival of the seven-spot, and by the decline of familiar native species, especially the transverse ladybug. I originally intended this book to be a call to arms against such alien invaders, and a sort of eulogy for the native ladybugs of our province. But as I became increasingly familiar with the situation, I started to find native species all around me, most of them right where I remembered them from my youth (I was a beetle collector as a school kid; a sort of person you just don't run into nowadays). It was clear that some of the old familiar ladybugs had become quite rare, but not all. Things have changed, but is change in itself a bad thing?

As a quasi-paleontologist (I have had a long working association with the Royal Tyrrell Museum), I am continually reminded that the one constant in nature is change. The history of Alberta since the end of the last ice age is a case in point. Most people assume that the "unusual" ice melted, and things became "normal," but in truth, the last 10,000 years have been characterized by fluc-

A winter ladybug climbs the leaf of a sage plant on the prairies.

tuations in climate, vegetation, and the nature of the landscape. To believe that there was a stable, "original" state of nature that existed before the arrival of European people is simply a mistake. All evidence suggests continual change. It was slow gradual change for the most part, but still change. Therefore, much of our common sense view of the world, and especially the rather stabilist approach we take toward the conservation of nature, is based more on myth and nostalgia than on ecology and biogeography. Making peace with ecological change may be difficult, but to me, it makes sense as the only real option.

There are two common responses to ecological change: you can resent it and fight against it or find it interesting and worthy of study. You may think the second option odd, but let me remind you that many modern ecologists believe that the arrival of new species is best viewed as a series of accidental experiments that test and clarify our views about ecology and evolution (especially when nature has rearranged itself without any extinctions taking place). To see it this way, one must also agree that dispersal is a normal part of the history of life, humans are a part of nature, human-assisted dispersal is not really different in kind from so-called natural dispersal, human-assisted dispersal is a predictable result of human dispersal itself, the term "native" has no real meaning in biology, there never was an original state of nature, there is no such thing as the balance of nature, and there is no such thing as "the way nature intended." These things may not be readily apparent to everyone reading

The nine-spot ladybug waves hello from a scurf pea plant on the margin of a sand blowout near Purple Springs, happy to be alive and well despite rumours to the contrary.

this book; however, to most biologists with a background in so-called basic or "conceptual" ecology (driven by the need to understand nature, as opposed to applied ecology driven by the need to control nature), they make perfect sense. In contrast, those who instantly label any and all introduced species as "alien invaders" or "threats to biodiversity" may be missing the real lessons that these newcomers can teach us about the way nature works.

Some people will no doubt accuse me of undermining the conservationists' cause (and perhaps coming across as an Alberta redneck) by my reluctance to demonize introduced species of ladybugs, but undermining conservation is the furthest thing from my mind. Instead, I am trying very hard to follow my conscience and remain as scientifically self-honest as I can. Conservation, to me, is about sustaining the living world around us—people included—not about returning to a mythical golden era when we, or our ancestors, or someone else's ancestors, were "in harmony with nature." The biggest threat to conservation I see is the loss of credibility that comes from "crying wolf," and I think that ladybugs provide a case in point. The second biggest threat to conservation is loss of habitat, not introduced species.

I dislike industrial development, urban sprawl, and the conversion of natural or semi-natural landscapes into cropland as much as the next naturalist, but my point here has nothing to do with these things—it is about the nature of the living world, and the need to make peace with inevitable change. The new ladybugs are here, more are likely on the way, and my job is to lay out for you a detailed picture of what we know, what we don't know, and what kind of sense we can make of these patterns from the perspective of biology.

It was a ladybug specialist, Theodozius Dobzhansky, who gave us one of science's most famous quotes: "Nothing in biology makes sense except in the light of evolution." Ladybugs are the dynamic products of evolution by natural selection, not off-the-shelf cogs in an environmental machine. What this means takes time and patience to unravel. Conservation biologists will tell you that their field is now in its "ecological–evolutionary" phase, in step with modern science, having outlived its so-called "romantic–transcendental" phase, tied as it was to religious and cultural biases, and rather outdated notions about the nature of the living world. I personally think that this announcement is premature and that many, if not most, people in the conservation arena are engaged in a hopeless struggle against change and toward a stabilist goal for the environment, loosely based on naïve notions of what things were like here some 300 years ago. However, as philosopher Daniel Dennett wrote, the evolutionary view bears "an unmistakable likeness to universal acid: it eats through just about every traditional concept and leaves in its wake a revolutionized world views, with most of the old landmarks still recognizable but transformed in fundamental ways." If evolution is about change (and it is), then change, in and of itself, cannot be the enemy in an evolutionary version of conservation biology. We are not living among the finished products of the evolutionary process—we are living within the currents of the process itself, but that is a much bigger topic for discussion and perhaps not one to hide in the preface of a ladybug book.

Gallery of the Lesser Ladybugs of Alberta

Micro Ladybug
Microweisia misella, p. 59

American Hairy Ladybug
Stethorus punctum, p. 61

Newcomer Hairy Ladybug
Stethorus punctillium, p. 62

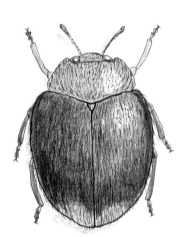

Mealybug Destroyer
Cryptolaemus montrouzieri, p. 63

Twice-stained Ladybug
Didion punctatum, p. 64
spotted form

Twice-stained Ladybug
Didion punctatum, p. 64
unspotted form

Angular Ladybug
Didion longulum, p. 65

Apicanus Ladybug
Scymnus apicanus, p. 66

Paracanus Ladybug
Scymnus paracanus, p. 67

Opaque Ladybug
Scymnus opaculus, p. 68

Fake Opaque Ladybug
Scymnus postpictus, p. 69
male

Fake Opaque Ladybug
Scymnus postpictus, p. 69
female

Carr's Ladybug
Scymnus carri, p. 70

Uncus Ladybug
Scymnus uncus, p. 71

Diamond City Ladybug
Scymnus aquilonarius, p. 72

XVIII   *Gallery of the Lesser Ladybugs of Alberta*

Lacustrine Ladybug
Scymnus lacustris, p. 73
male

Lacustrine Ladybug
Scymnus lacustris, p. 73
female

Ornate Ladybug
Nephus ornatus, p. 74

Farmer's Ladybug
Nephus georgei, p. 75

Sordid Ladybug
Nephus sordidus, p. 76

Tinytan Ladybug
Selvadius nunenmacheri, p. 77

Mimic Ladybug
Hyperaspidius mimus. p. 78

Vittate Ladybug
Hyperaspidius vittigerus, p. 79

Well-marked Ladybug
Hyperaspidius insignis, p. 80

Gallery of the Lesser Ladybugs of Alberta   XIX

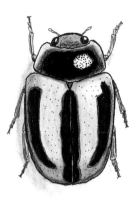

Hercules Ladybug
Hyperaspidius hercules, p. 81

Unnamed Ladybug
Hyperaspidius sp., p. 82

Convivial Ladybug
Hyperaspis conviva, p. 83

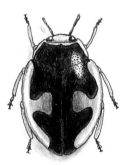

Lugubrius Ladybug
Hyperaspis lugubris, p. 84

Lateral Ladybug
Hyperaspis lateralis, p. 85
light form

Lateral Ladybug
Hyperaspis lateralis, p. 85
dark form

Fastidious Ladybug
Hyperaspis fastidiosa, p. 86

Curved Ladybug
Hyperaspis inflexa, p. 87
light form

Curved Ladybug
Hyperaspis inflexa, p. 87
dark form

Postica Ladybug
Hyperaspis postica, p. 88

Oregon Ladybug
Hyperaspis oregona, p. 89
light form

Oregon Ladybug
Hyperaspis oregona, p. 89
dark form

Blotch-backed Ladybug
Hyperaspis disconotata, p. 90

Undulate Ladybug
Hyperaspis undulata, p. 91

Poorly-known Ladybug
Hyperaspis consimilis, p. 92

Four-streaked Ladybug
Hyperaspis quadrivittata, p. 93

Jasper Ladybug
Hyperaspis jasperensis, p. 94

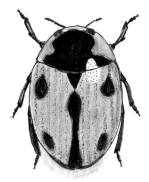

Pale Anthill Ladybug
Brachiacantha albifrons, p. 95

Gallery of the Lesser Ladybugs of Alberta   XXI

Ursine Anthill Ladybug
Brachiacantha ursine, p. 96

Underside of abdomen showing complete postcoxal arcs.

Underside of abdomen showing incomplete postcoxal arcs.

Gallery of the Larger Ladybugs of Alberta

Winter Ladubug
Brumoides septentrionis, p. 101

Round Black Ladybug
Exochomus aethiops, p. 102

Twice-stabbed Ladybug
Chilocorus stigma, p. 103

Once-squashed Ladybug
Chilocorus hexacyclus, p. 104

Snow Ladybug
Coccidula lepida, p. 106

Marsh Ladybug
Anisosticta bitriangularis, p. 108

Episcopalian Ladybug	Thirteen-spot Ladybug	American Ladybug
Macronaemia episcopalis, p. 109	*Hippodamia tredecimpunctata*, p. 111	*Hippodamia americana*, p. 113

Waterside Ladybug	Parenthesis Ladybug	Parenthesis Ladybug
Hippodamia falcigera, p. 114	*Hippodamia parenthesis*, p. 115 heavily marked	*Hippodamia parenthesis*, p. 115 lightly marked

Expurgate Ladybug	Expurgate Ladybug	Five-spot Ladybug
Hippodamia expurgata, p. 117 heavily marked	*Hippodamia expurgata*, p. 117 lightly marked	*Hippodamia quinquesignata*, p. heavily marked

Gallery of the Larger Ladybugs of Alberta

Five-spot Ladybug
Hippodamia quinquesignata, p. 118
lightly marked

Glacial Ladybug
Hippodamia glacialis, p. 119
heavily marked

Glacial Ladybug
Hippodamia glacialis, p. 119
lightly marked

Sorrowful Ladybug
Hippodamia moesta, p. 121

Colonel Casey's Ladybug
Hippodamia caseyi, p. 122

Convergent Ladybug
Hippodamia convergens, p. 123

Boulder Ladybug
Hippodamia oregonensis, p. 124

Sinuata Ladybug
Hippodamia sinuate, p. 125

Gallery of the Larger Ladybugs of Alberta   XXV

Flying Saucer Ladybug
Anatis rathvoni, p. 126

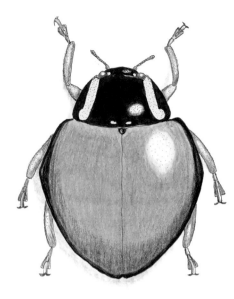

Large Orange Ladybug
Anatis lecontei, p. 127

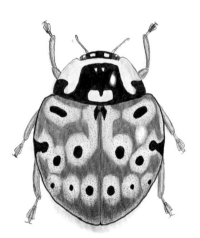

American Eyespot Ladybug
Anatis mali, p. 128

Streaked Ladybug
Myzia pullata, p. 129

Subvittate Ladybug
Myzia subvittata, p. 130

Polkadot Ladybug
Calvia quatuordecimguttata, p. 131
pink form

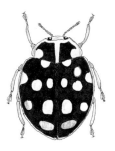

Polkadot Ladybug
Calvia quatuordecimguttata, p. 131
black form

Polkadot Ladybug
Calvia quatuordecimguttata, p. 131
pale form

Two-spot Ladybug
Adalia bipunctata, p. 132
two-spotted form

Two-spot Ladybug
Adalia bipunctata, p. 132
four-spotted form

Two-spot Ladybug
Adalia bipunctata, p. 132
melanic form

Two-spot Ladybug
Adalia bipunctata, p. 132
banded form

Two-spot Ladybug
Adalia bipunctata, p. 132
spotless form

Gallery of the Larger Ladybugs of Alberta XXVII

Two-spot Ladybug
Adalia bipunctata, p. 132
many-spotted form

Two-spot Ladybug
Adalia bipunctata, p. 132
melanic form

Three-banded Ladybug
Coccinella trifasciata, p. 134

Transverse Ladybug
Coccinella transversoguttata,
p. 135

Seven-spot Ladybug
Coccinella septempunctata, p. 137

Nine-spot Ladybug
Coccinella novemnotata, p. 139

High-country Ladybug
Coccinella alta, p. 141

Tamarack Ladybug
Coccinella monticola, p. 144

Hieroglyphic Ladybug
Coccinella hieroglyphica, p. 14

Polished Ladybug
Cycloneda polita, p. 147

Halloween Ladybug
Harmonia axyridris, p. 148
typical form

Halloween Ladybug
Harmonia axyridris, p. 148
sparsely-marked form

Halloween Ladybug
Harmonia axyridris, p. 148
melanic form

Painted Ladybug
Mulsantina picta, p. 151

Hudsonian Ladybug
Mulsantina hudsonica, p. 152

Wee-tiny Ladybug
Psyllobora vigintimaculata, p. 153

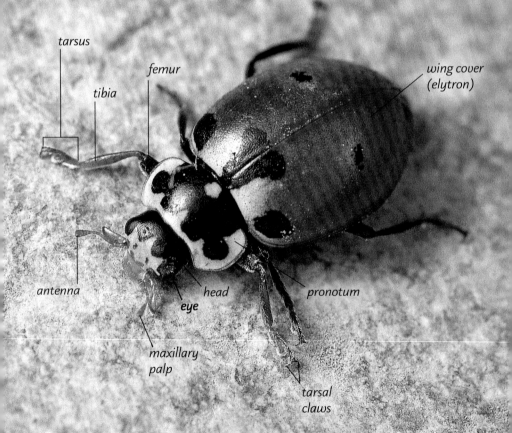
Main features of an adult ladybug.

1
What is a Ladybug?

Ladybugs are some of our most familiar insects, to be sure. Just about everyone loves them, at least here in Alberta, and I am delighted to say that this is the first ladybug field guide for any part of the Western Hemisphere. It would be nice if there were only a few different types of ladybugs in Alberta, but the truth is that there are 75 or more different species, some of which are quite variable in their appearance. If you are up to the challenge, ladybug identification is reasonably easy and fun, and I'm delighted to help Albertans pioneer this aspect of nature appreciation.

So let's begin with the simple question: What is a ladybug? Technically, ladybugs are members of a single family of beetles, the Coccinellidae. Since they are beetles, some people dislike the name "ladybug;" the so-called "true bugs" belong in the insect order Hemiptera, and beetles are members of the Coleoptera. For this reason, you will often see them called "ladybirds," "ladybird beetles," or "lady beetles." To me, it is silly to try to re-educate Albertans, who almost universally call them ladybugs. It's about as hopeless as getting people to say "Richardson's ground squirrel" instead of "gopher." Besides, I wrote an entire book on damselflies knowing perfectly well that they were not flies, so I don't feel bad about writing on ladybugs knowing they are not bugs.

The "lady" part of their name, by the way, refers to the Virgin Mary, and they are in that sense actually "Our Lady's Bugs." The red colour of many ladybugs reminded some early Europeans of the red robes of the Virgin Mary, and this is said to be the origin of the name. "Coccineous" is a term meaning "berry red" and is of course the Greek source from which the name Coccinellidae is derived. One prominent entomologist, Willis S. Blatchley, attempted to promote the name "plant louse beetles" alongside "lady bugs" and "lady beetles," but, thankfully, that idea never caught on.

What should we call ourselves, those of us who study ladybugs? I made the mistake, not long ago, of asking a man standing next to me in the poster section of a scientific meeting, if he was "a ladybug guy." He looked at me quizzically, and I rephrased it—"Oh, I mean... are you a coccinellidologist?" At this point, his eyes lit up, and he turned toward me ready to chat. It was John Sloggett, a well-known British ladybird man, and I was pleased to meet him. Personally, I like the term "ladybugster," but I'm not sure it will catch on. How about "ladybug enthusiast" or "ladybug lover?" They all have their shortcomings, but as long as we can recognize each other as kindred spirits, we'll be all right.

As beetles, ladybugs have forewings that are hardened into wing covers, or "elytra," that meet in a straight line down the back of the beetle, and cover the folded hind wings, which are used for flight. Most ladybugs are less than 10 mm long, and some are less than 2 mm in total length. Ladybugs generally have rounded or oval bodies, relatively short legs, and relatively short antennae with small but obvious clubs at their tips. Entomology texts will tell you to count the segments of the tarsus or "foot" of a ladybug, and look for the hidden third segment at the base of the fourth segment in order to confirm that you do indeed have a ladybug in front of you. I find this difficult to see at times, even with a microscope, since it is indeed hidden, and the fact that some ladybugs (in the genus *Nephus*) have only three tarsal segments has convinced me to look for their other distinguishing features in times of confusion. If you know your beetles reasonably well, you will find that all ladybugs do indeed possess "the ladybug look."

The evolutionary place of ladybugs among the beetles is fairly well understood, and the ladybugs comprise one family among many in a large group of quite obscure small beetles called the "Cucujoidea." Their closest relatives include the families Latridiidae (the minute brown scavenger beetles), and Endomychidae (the handsome fungus beetles)—creatures that only specialists recognize, for the most part. Surely, on this branch ("the cerylonid series") of the massive family tree of beetles, the ladybugs are the most famous, most obvious, and best-loved members.

Plenty of red and black, or orange and black creatures look like ladybugs, so don't be fooled! (As well, please be sure that what you are looking at is a ladybug, before you write to me and tell me disdainfully that I "missed" a species!) Be especially wary of some leaf beetles (in the family Chrysomelidae), which generally have longer antennae and a different look in their eye. There are also red and black stink bugs, most of which are ladybug-like only when they are young and lack wings. Another true bug, the superb plant bug (*Adelphocoris superbus*) is quite ladybuggish as a winged adult. When searching for the tiniest of ladybugs, even careful beetlers often pick up large numbers of "shining flower beetles," in the family Phalacridae, small smooth beetles with clubbed, elongate antennae. (Despite my having sorted through hundreds of phalacrids in search of tiny ladybugs, the official checklist of Alberta shining flower beetles includes a grand total of zero species, so this is obviously a place where discoveries might be made, if ladybugs themselves seem too mainstream for you.)

Some ladybug lookalikes, *top left: superb plant bug* (Adelphocoris superba), *top right: red turnip beetle* (Entomoscelis americana), *bottom left: four-spot leaf beetle* (Chrysomela quadriguttata), *bottom right: hister beetle* (Spilodiscus instratus).

Some more ladybug lookalikes, top left: *willow leaf beetle* (Chrysomela scripta), top right: *shining flower beetle* (Phalacrus sp.), bottom left: *black-footed tortoise beetle* (Jonthonota nigripes), bottom right: *immature two-spotted stink bug* (Perillus bioculatus).

In this book, I have tried to make ladybug identification both easy and accurate. Most species can be identified simply by examining the colour pattern on the wing covers and the pronotum (the shield on the top of the beetle, between the head and the elytra). Most of the important features are on the dorsal (top) side of the beetle, but on occasion you will need to look at the ventral (under) side as well. Perhaps the most important features to see are the so-called post-coxal arcs on some of the lesser ladybugs: thin lines on the first segment of the abdomen behind the base of the hind legs. For this, you will need a "dissecting" or "stereo" microscope (or perhaps a hand lens and a steady hand) that magnifies at least 20 times. This is also the only instance where you might need dead specimens. Most ladybugs can be identified while still alive or from good macro photographs.

The colouration of typical ladybugs—black with yellow, orange, or red—is a warning to predators that they taste bad (and I should add here that I performed some quick-and-dirty tests for ultraviolet reflections using a digital camera and a UV-cutoff filter, but found nothing that looked like ultraviolet patterning in ladybug specimens). Ladybugs protect themselves with compounds in their blood called alkaloids. When they are threatened, as for example in a bird bill or between the tips of human fingers, they perform "reflex-bleeding" in which the blood (or "haemolymph") escapes from joints in the legs, causing no long-term harm to the ladybugs but causing the predator to release the beetle. This is why you see orange liquid on your fingers when you pick up a ladybug too firmly. Many of these alkaloids have been named for the ladybugs from which they were derived, with names such as coccineline, hippodamine, and adaline, named for the genera *Coccinella*, *Hippodamia*, and *Adalia* respectively.

Not all ladybugs taste equally bad. Since the sense of taste in birds is immensely difficult to investigate (involving careful experiments with naïve captive reared birds, among other things), I often taste the reflex blood of ladybugs myself. This is not nearly as weird as it sounds, but I'll admit it does sound weird. If there were any diseases or parasites I could catch, I wouldn't do it, believe me. The procedure is simple: I place the ladybug on my tongue, press them gently against the roof of my mouth, remove them unharmed, and swirl the saliva around my mouth. I then take notes. I have determined that, to my taste at least, the worst ladybugs are the seven-spot, the transverse, the twice-stabbed, the Halloween, and the ash-gray (*Olla v-nigrum*). All of these are bitter tasting and put me off my appetite. By contrast, most of the *Hippodamia*, as well as the two-spot taste mildly musty, something like the smell of two-week-old salad, or dirty socks. The only lesser ladybug I have tasted is the lateral ladybug—my field notes read "subtle, bitter, metallic, and recognizably a ladybug alkyloid." The absolute worst is a relative of the twice-stabbed ladybug, *Chilocorus cacti*, which I tasted in Florida—it nauseated me for a full eleven hours.

It may be that less-distasteful ladybugs mimic the more distasteful species. In Britain there is evidence for two "mimicry rings," both of which involve

Top: A two-spot ladybug reflex bleeding—note the drop of orange blood on the right edge of the pronotum. Bottom: the worst tasting ladybug I have ever encountered, Chilocorus cacti from Florida.

a number of species. There are light-coloured beetles in one ring and dark-coloured beetles in another, and some polymorphic species (species with more than one colour form among their members) participate in both. This has not been studied here in North America, but it seems likely that mimicry is involved here too. It is also likely that some of the leaf beetles and stink bugs that are often mistaken for ladybugs are involved in these mimicry associations as well.

One of the great mysteries of ladybug colouration is simply this: why, if ladybugs are warningly coloured, do some species occur in a variety of colour patterns? Why not show the predators just one pattern, so the predators can learn more effectively? Apparently, there are two aspects of mimicry theory that might help here. First, it might be that predators don't distinguish among the colour morphs and easily learn to avoid all ladybugs regardless of pattern. I'll let you look at the illustrations and decide for yourself if this is reasonable, but to me it seems unlikely, especially when you consider the diversity of other small colourful insects out there. Second, there may be mimicry involved, and the polymorphism may be a distasteful species' way of evolving away from its less distasteful mimics (which reduce the effectiveness of the defence). Since some individuals die while the predator is being educated by the distasteful ladybug, perhaps evolution favours those who look less like the mimics. When there are mimics around, educating the predators is not as efficient, and the warning colouration is not as effective at saving ladybug lives. Do I believe this? Well, sort of—but it sounds improbable to me.

Colour polymorphism may not have much to do with mimicry at all. For polymorphic species, it has also been suggested that dark colour ("melanism") is related to increased humidity, aridity, industrial pollution, and the like, but in a complex and poorly understood way. In warmer areas where ladybugs are out and about year-round, dark forms of some species do best in summer, whereas lighter beetles do best in winter. One good example is the two-spot ladybug, and this species includes both light (red) and dark (mostly black) individuals both here and in Europe. As well, the genetics of the two-spot are quite well understood, making this a popular subject for this sort of research.

In some studies, females preferred one colour of male to the other, and it seems that females also seem to prefer larger and more active males and to recognize them in part by their smell. When the characteristics of living things come from the effects of the opposite sex choosing some sorts of mates over others, we speak in terms of "sexual selection," and in my conversations with other ladybugsters this is the explanation that seems to make most sense when explaining polymorphism.

There is another pattern here that makes little sense to me so far. All of our polymorphic ladybugs live in trees (the two-spot, polkadot, and Halloween), and all of the low-vegetation species are generally uniform in their appearance. Perhaps this has to do with greater threat from bird predators in trees. Perhaps

Two very dusty, but still lusty, two-spot ladybugs—note that the female is larger, and that neither of these individuals has only two spots.

it is related to the presence of the extremely bad-tasting twice-stabbed ladybug in trees but not in low vegetation. Or perhaps there is an even more subtle explanation. Among mimicry biologists, it has become increasingly important to look at the background context of colour patterns, and perhaps there is something about the "look" of a ladybug in a tree, as opposed to on a herb, that promotes the origin of more than one colour pattern within a species. Who knows, but I do hope that some patient, well-funded person is able to pursue this as part of their research sometime soon.

Speaking of which, it is sometimes relatively easy to tell male and female ladybugs apart. Males are generally smaller, while females are generally larger—the usual pattern in most insects. As well, the males may have a broader indentation on the trailing edge of the last or second-last abdominal segment. Among the lesser ladybugs, it is common for males to have more yellow on their face than the females. However, in most common species of ladybugs I can't see any obvious sexual differences, and I doubt it is easy to tell them apart.

Of course, ladybugs are also prominent in Western culture, and other human cultures as well. We have all heard the poem "Ladybug, ladybug, fly away home…" and we have all been told that the ladybug is our friend. Gardeners, especially, like to think of ladybugs as just the right sort of bugs

Variation in two-spot ladybugs: these are all two-spots, as are those with only two spots. This has caused taxonomic confusion in the past, and makes identification difficult even today. Hint—they are most similar on the pronotum.

Top: Anatis mali, *the American eyespot ladybug on its namesake,* Malus *the apple. Bottom: Three colour forms of the Halloween ladybug, photographed in Washington, D.C.*

for the garden, since ladybugs are small, cute, and eat pestiferous aphids. I am absolutely amazed at home many gardening shops use ladybugs (and typically the seven-spot ladybug) in their advertising logos. How odd, considering that the object of gardening is to grow plants, not insects! Add the sheer number of ladybug decorations and toys for sale in garden shops and nature centres, and the message is clear: we love ladybugs and they can do no wrong.

In other parts of North America where the Halloween ladybug is abundant, people have lost much of their affection for these beetles, since this newly arrived species often congregates in houses and causes a noticeable odour, an unmistakable insect presence through the winter, and the occasional bite. In these areas, it is as if the ladybugs have "gone bad," and entomologists are called upon often to explain how our favourite beetles could betray their human friends so badly. I have also heard people mistakenly refer to Haloween ladybugs as "those things that look like ladybugs but really aren't."

Of course, there are plenty of oddball beliefs about ladybugs out there as well. It should be clear to you that at least two popular beliefs about ladybugs are not true. First, the number of spots on a ladybug does not tell you how old it is, or how harsh the coming winter will be. Second, small ladybugs do not grow into big ladybugs. Instead, all ladybugs are adult beetles. Each individual ladybug was originally an egg, then a larva, then a pupa, and finally an adult beetle, and next I will describe the life of a ladybug in more detail.

Aphid-eaters, such as this five-spot/glacial ladybug, will eat pollen and other alternative foods when aphids are scarce.

2

The Life of a Ladybug

All ladybugs begin life as eggs, and the eggs of ladybugs are small, usually yellow or orange, and typically laid in clumps on a plant where the food of the larvae can be found. Interestingly, most ladybug egg clusters include a number of infertile eggs, and this is not an accident. The newly hatched larvae eat the infertile or "trophic" eggs, and for many baby ladybugs this is their first meal. Jennifer Perry, a graduate student at Simon Fraser University in British Columbia, found that when there were few aphids present the mothers laid more trophic eggs, presumably helping the new larvae along through tough times. Also, like many newly hatched insects, newly hatched ladybug larvae also eat their own eggshells.

The larvae of ladybugs are elongate, six-legged creatures with clearly defined body segments. Most people find them far less cute than the adults. As the larvae grow, they pass through four "instars" between which they shed their skins. At warmer temperatures and with more food, they grow more quickly, and their rate of growth can be remarkably fast. For example, Cesar Rodriguez-Saona and Jeffrey Miller found that the time it took convergent ladybugs to develop from egg to adult was a mere 14 days at 30 °C but a full 51 days at 18 °C. Moderation is the key, however, and adult weight and wing size was greatest among beetles that developed at an optimum temperature of 22 °C. J.Y. Xia and colleagues found that seven-spots also developed most quickly at high temperature but that egg to adult survival was highest at a more reasonable 25 °C, as was the number of eggs produced by adult females.

These sorts of studies allow entomologists to calculate the number of "degree-days" needed for development from egg to adulthood. To calculate degree-days, you take the average temperature each day, and subtract from it the minimum temperature at which the ladybugs can grow, the so-called "developmental threshold" (which is generally somewhere between 12 °C and 14 °C). For the convergent ladybug, the study by Rodriguez-Saona and Miller estimated that 231 degree-days are required for complete development, above a threshold of 13.6 °C. To increase their accumulation of degree-days, ladybugs

Top: A batch of ladybug eggs, beside a naïve young aphid, doomed to be devoured once the eggs hatch. Bottom: A fat, velvety, "cigar-shaped" seven-spot ladybug larva, fully grown.

often forage most heavily on the sunlit side of trees and bushes. I have noticed this with two-spot and seven-spot ladybugs feeding on aphids on caragana.

Because many ladybug larvae eat economically important pest aphids and scale insects, the feeding habits of ladybug larvae have been well studied. Aphids also attract other predators, such as lacewing larvae and the larvae of some hover flies, so aphid-rich plants can be dangerous places and not just for aphids. Ladybug larvae can also be cannibalistic, eating larvae and eggs of their own species or others.

Cannibalistic Halloween ladybug larvae are more likely to survive and develop quickly than their non-cannibalistic kinfolk. At least that's what William Snyder and his colleagues found, as well as determining that, if the larvae fed on other larvae that in turn fed on high-quality aphids (those without many toxic chemicals in their bodies), the benefits to the cannibals were even greater. If the larvae were fed only on poor- and intermediate-quality aphids, only those that fed on intermediate aphids (and/or larvae that themselves fed on intermediate quality aphids) survived to become adults. It seems likely that the cannibal larvae take advantage of the fact that their victims have already detoxified the lower quality aphids, but there are clearly still things here to be learned.

Other sorts of insect eggs can also serve as ladybug food. For example, one study found that twelve-spot ladybugs (*Coleomegilla maculata*) could develop on Colorado potato beetle (*Leptinotarsa decemlineata*) eggs, but not all the larvae made it to adulthood. This was the work of Joseph Munyaneza and John Obrycki. Mpho Phoofolo (the man with the best name in ladybug science, in my opinion) and John Obrycki found that twelve-spot and Halloween ladybugs could develop to adulthood eating nothing but lacewing eggs, but that seven-spot larvae could not. In turn, lacewing larvae (*Chrysoperla carnea*) could develop on nothing but twelve-spot eggs. As well, both the lacewing and ladybug larvae eat each other, although mostly this means the lacewings eat the ladybugs.

The influence of plant defence chemicals on the development of two-spot ladybug larvae feeding on peach-potato aphids (*Myzus persicae*) was the subject of a study by Frederick Francis and his colleagues. They reared aphids on canola (a plant with low levels of "glucosinolate" defence chemicals), mustard (which produces high levels of "GSL"), and beans (which have no GSL). In general, oddly enough, the aphids did best on canola and mustard and less well on beans. The canola and mustard-fed aphids produced faster growing larvae and larger ladybugs. However, mustard-fed aphids resulted in smaller ladybug eggs, fewer eggs, and fewer viable eggs. Like so much of ladybug ecology, I'm sorry to report, it makes no obvious intuitive sense at all—nature is so wonderfully but frustratingly complicated. Keep this in mind when we come back to the task of unraveling the effects of introduced species.

In this book, I have not attempted to provide a guide to ladybug larva identification, but it is likely that all of our species could eventually be told apart as larvae. There is a key to the larvae that takes them to the genus level, written by B.E. Rees and colleagues. Bob Gordon and Natalia Vandenberg have also produced a fine paper on the larvae of *Coccinella* ladybugs and for now that is our best guide. In general terms, the common larvae of *Coccinella* ladybugs can be separated from *Hippodamia* by the dark markings on the pronotal shield behind the head, which are connected at the hind edge in *Hippodamia*, and around the middle in *Coccinella*. Two-spot larvae have shorter legs and a more oval body. *Calvia* and *Myzia* larvae have a point or "appendix" on the top of the ninth abdominal segment. "Other genera," according to Gordon and Vandenberg, "differ conspicuously in the form of the body armature, particularly the dorsal setose processes, and are unlikely to be confused with *Coccinella*."

When a ladybug larva finishes its development, it generally leaves the plant on which it was hunting, and searches for a place to become a pupa. Ladybug pupae are usually formed in the open, attached to a surface by the tip of the abdomen. They are generally warningly coloured, in orange or yellow and black or brown, and they can wiggle a bit, presumably to scare off potential parasites. Colour may vary among pupae of a single species, and as with adults, darker pupae are apparently the result of increased humidity or lower temperatures, at least in part. Among the lesser ladybugs, the last larval skin covers the pupa, but this is not the case among the larger ladybugs.

After relatively short period (days or weeks, but not months), the pupa "ecloses" to become the adult ladybug. At first, the beetle is soft with pale colours. This is the teneral phase of the life history, during which chemical changes in the cuticle (the outer surface or exoskeleton) of the beetle both harden and darken the body. Within a single species, there can be quite a bit of variation in final body size, but males are generally smaller than the egg-bearing females.

Once ladybugs are past their teneral phase, they proceed with the business of feeding and finding mates. Males can mate more-or-less right away, since the production of sperm begins while they are in the pupa. Females can also mate immediately and store sperm for later use in a structure called the spermatheca. One mating is all that a female needs to fertilize all of her eggs for life, but most females mate multiple times nonetheless. In some species, including our most familiar ladybugs such as the seven-spot, the female can lay hundreds of eggs if she survives long enough to do so.

Unless indicated otherwise in the text that follows, it is generally the case that ladybugs eat aphids. Aphids are small sucking bugs in a number of families that in turn fall into a broader group called the Aphidomorpha. I wish I could recognize different species of aphids in the field, but I'm having enough trouble learning the ladybugs. Most aphids sit on the stems or leaves of plants, and

Top left: The voracious, aphid-eating larva of a lacewing, with long piercing mandibles. Top right: The elegant and bad-smelling adult green lacewing. Bottom: A two-spot ladybug at an aphid smorgasbord, eating its fill.

Top left: A two-spot ladybug larva about to pupate, having attached the tip of its abdomen to a branch. Top right: The newly formed pupa of a seven-spot ladybug, pale yellow in colour.

Botton left: Side view of a seven-spot ladybug pupa—note the warning colours, and the crumpled larval skin at the base. Bottom right: The empty pupal skin of a twice-stabbed ladybug, hanging by a thread below the spiny larval skin in which it spent the pupal stage.

suck the sugary juices from the phloem of the plant. In order to eat an aphid, a ladybug or ladybug larva simply walks up to it, and devours it. Aphids will kick to fight back, and they do produce defence chemicals; however, in most instances this is of little help in when confronted with a hungry ladybug.

Most aphids over-winter as pregnant adult females that emerge in spring and start colonies on plants (although many also overwinter as eggs). They go through a number of largely wingless parthenogenetic generations (females giving live birth to more females, in the absence of males) and then in the fall produce both winged females and winged males. Thus, aphid numbers increase very quickly over the course of the growing season, and ladybugs have in part synchronized their own life cycles to the annual waves of aphids, in an evolutionary sense. This also means that aphid numbers can almost always stay ahead of ladybug numbers, which is why ladybugs never manage to eat all the aphids.

Ladybugs that feed on aphids will also take a variety of alternative foods, especially during times when aphids are scarce. These alternative foods include other sorts of insects, insect eggs, sap, and even pollen. Thus, it is perfectly normal to find ladybugs in dandelions in the spring, or at sap-flow wounds on trees such as birch and maple. When I put out a bait of beer, rum, molasses and brown sugar for night-flying moths, I sometimes get a ladybug or two sipping on the edge of the bait patch, painted on tree bark. The polkadot ladybug seems especially easy to attract to bait, and I once had a polkadot ladybug come to a half eaten bowl of carrot and oatmeal baby food while we were on a family camping trip at Lac La Biche.

In many instances (in lawns, sedges, cornfields, and wolf willows, for example), you will encounter large populations of ladybugs but no aphids. Every time this happens, my first thought is that I have missed the aphids, but in most instances there are leaf-hoppers and plant-hoppers present, and these are the obvious prey. It's embarrassing, in retrospect, to admit that I have always been too busy to sit down with close-focusing optics and actually watch to see if the ladybugs actually eat the leaf or planthoppers.

Because some plants predictably harbour more and better quality aphids than others, it is easy to know where to look for large concentrations of ladybugs. Among my own favourite searching spots, I include the following trees: birches (especially weeping birch), lone jack pine, spruce or lodgepole trees, set apart from clumps and forests, manitoba maple trees ("boxelders"), mountain ash trees (where ladybugs may be aestivating in the berry clusters), and balsam poplar seedlings (aspen are poor). The best bushes are wolf willow, sage, caragana (where aphids feed on the pods), cherry and alder (for some lesser ladybugs on bark) and Saskatoon (with leaf mildew for wee-tiny ladybugs). I also habitually search scraggly grasses on dune ridges and pediment slopes, thistles, alfalfa fields, sedges, and honeysuckle vines with wooly aphids. Snowberry patches can also be worth searching.

Adult ladybugs live from a month or so to up to three years. Of our species, the longest-lived is the Halloween, which has been documented at three years of age. Some other sorts of beetles also live this long, but among adult beetles, this sort of longevity is certainly the exception not the rule. Seven-spot ladybugs can live for a second year as an adult, and it is tempting to suggest that longevity is in part responsible for the success of these two recently introduced species in North America. Add to this the fact that most of our ladybugs go through two breeding generations per year, and the potential for their populations to rapidly increase becomes readily apparent.

Adult ladybugs are good fliers, and they fly almost exclusively by day. Some of us, who run light traps for moths and other night-flying insects, have records of the occasional ladybug in our traps, but night flights are not a regular part of their daily life. The reason ladybugs fly so much is that they rarely settle in to a particular patch of aphids for very long. Aphid colonies boom and bust like oilfields in rural Alberta, and the ladybugs are the equivalent of rig workers following the jobs. Many female insects, while producing eggs, lose much of their flight musculature and are temporarily unable to fly, but this is not the case among ladybugs—they need to remain mobile in order to survive.

Female ladybugs lay eggs more readily in the presence of dense populations of aphids, but they will also lay eggs in places where there are no aphids at all. Egg laying can occur more or less throughout the season, but in almost all of our species, it is the adult beetle that spends the winter not the egg, larva, or pupa.

Hibernating congregations of ladybugs can be quite dramatic, and the most obvious of these are found in leaf litter (for example, seven-spot ladybugs) or in houses (two-spot or Halloween ladybugs). On hilltops, especially in the foothills, you might also find hibernating clusters of various *Hippodamia* ladybugs. This phenomenon is especially well known along the West Coast, where convergent ladybugs are collected for resale to gardeners, a laudable idea that unfortunately doesn't work.

Let me explain. The reason most entomologists recommend against ladybug releases is that the ladybugs still feel the need to make a long-distance flight when they leave what they think is their hibernaculum. They simply do not stick around to control *your* aphids, when they are faced with a wild and wonderful spring world *filled* with all the aphids they could dream of. For example, at a recent nature festival in the United States, I was asked to preside over the release of ladybugs in a town garden. I'm sure the organizers were discouraged to hear me explain to the children that the ladybugs were more than likely going to spend the next few days eating aphids on weeds in roadside ditches and along railway lines. Sure enough, 90% of the beetles took wing the moment their screen cage was opened, but I was especially saddened by all of the trampled ladybugs left on the ground when we were finished, and in general it is clear to me that ladybug harvesting does more harm than good.

A seven-spot ladybug just after emerging from its pupa, still pale in colour and soft.

So why do we persist in believing that ladybugs are "the gardener's friends"? Perhaps it has to do with the history of ladybugs as alternatives to pesticides. Way back in 1889, a type of ladybug called the vedalia beetle (*Rodolia cardinalis*) was used with great success in California, to control a pest called cottony cushion scale on citrus trees. An inexpensive, environmentally clean technique thereby saved millions of dollars, and raised the hopes of entomologists everywhere for "biocontrol." This sort of success has never been repeated with ladybugs, a fact that has never quite made it into the public consciousness. Entomologists, by the way, refer to this as the "ladybird fantasy period," admitting up front that it is likely never to return.

John Obrycki and Timothy Kring recently reviewed the use of ladybugs in biological control. They found that, in order to encourage ladybugs, you need to reduce pesticides and provide refuge for the beetles. The rate of establishment of newly introduced ladybugs is low (less than the average for all biocontrol organisms in North America) and, in general, ineffective species are used because they are easy to collect at their overwintering aggregations. Surprisingly, the efficiency of ladybugs as predators is quite poorly understood (in part because they are highly mobile and eat a variety of things), and they are

Big, green, juicy aphids—the bread and butter of the larger ladybugs.

rarely surveyed properly in biocontrol studies. Ladybugs sometimes eat other biocontrol creatures or disrupt their positive effects. For example, convergent ladybugs collected in California have been shown to have no effect on aphids where they are released and instead may spread ladybug parasites and plant diseases (in this instance, a fungus of dogwood bushes). Furthermore, as I will discuss in the next chapter, introduced ladybugs can have detrimental effects on their native relatives.

In general, aphid-eating ladybugs cannot keep up with increasing populations of aphids each year, since aphids are so much better at reproducing. From the ladybug's evolutionary perspective, this is probably a good thing, since it ensures that each ladybug has a good chance of finding aphids to eat. However, some people just won't leave nature to its own devices. There is talk of attempting to breed wingless strains of two-spot and Halloween ladybugs (that will be forced to stay in one place and control the aphids in that locality), and possibly pesticide-resistant ladybugs as well (so you can throw both pesticides and ladybugs on your crop simultaneously). These developments are still theoretical, but while we wait to see if this idea materializes, it is interesting to note that the application of "food sprays" (sugar and protein, and something called "artificial honeydew") has been shown to increase ladybug populations in crops, without adding toxins to the scene.

Returning to the subject of ladybug aggregations, if you find a ladybug in your house you can feed it bits of banana, and liver-flavoured catfood. In one study, Irene Geoghegan and her colleagues actually used magnetic resonance microimaging to show that ladybugs on an artificial diet built up a mass of nutrient in their bodies, as if preparing for hibernation. So perhaps giving the beetle some cat food and putting it back in a cool corner might be a nice thing to do. Most ladybugs will not, however, live and breed without feeding on aphids, so long-term captive breeding of ladybugs is a difficult thing, requiring long-term breeding of aphids on the side.

Alberta is not particularly well known for ladybug aggregations, but back in the early 1980s, A.M. Harper and C.E. Lilly reported huge clusters of five-spot ladybugs at various places along the eastern slopes and in the Porcupine Hills, generally between 1250 and 2439 metres in elevation. The clusters (or "colonies") were found on west-facing slopes under debris and rocks, which no doubt helped insulate the beetles, especially under snow. These same ladybugs were thought to feed down on the prairies in summer, on pea and grain aphids in cropland, and then return to the high country in the fall. With that thought as motivation, these two crop pest entomologists measured the cold tolerance of the beetles; they found that the beetles developed cold hardiness in the fall and that their resistance to cold was greatest in mid-winter. In dormant ladybugs, the gut is free of food, and the fat body grows to serve as a winter reserve. Then, during a typical spring, the beetles sun themselves during warm days, return to the aggregations each night, and mate at the aggregations before dispersing downhill. Cold hardiness (the result of glycogen in the blood that protects the ladybugs down to about $-30\,°C$) is lost in the spring, and Harper and Lilly thought that a late spring cold snap could result in terrible mortality at these colonies. Their measured overwintering mortality in the colonies varied from 20% to 70%. They also found Colonel Casey's ladybug in small numbers in a few sites (Porcupine Hills and Turtle Mountain). The Harper and Lily study is clearly the best and possibly the only relevant work on ladybug hibernation at Albertan winter temperatures, and it would be interesting if further research was done on this phenomenon.

Why do some ladybugs aggregate for the winter? It seems to be a means of increasing the effectiveness of their warning colouration and defence chemicals. Predators that might eat a single, isolated ladybug might not have the stomach for a whole gut full. On the other hand, Karolis Bagdonas has told me that in Wyoming he has found grizzly bear droppings that contained nothing but the undigested remains of thousands of *Hippodamia* ladybugs. The bears typically, and famously, feed on winter aggregations of cutworm moths and occasionally they fill up on ladybugs as well.

Where ladybugs congregate in houses and buildings for the winter, the smell can be obvious to people ("penetratingly noticeable to man" according to Ivo Hodek and Alois Honěk), and this smell may be the means by which ladybugs find their winter colonies from year to year. In other words, it may be an "aggregation pheromone," although our understanding of such chemicals in ladybugs is still in its infancy

Some ladybugs will also enter a period of summer dormancy called aestivation, in order to sit out the period when aphid numbers are low. R.D. McMullen found that nine-spot ladybugs in California went into both aestivation (the spring generation of adults) and hibernation (the summer generation adults) in response to changes in day length during the first week of their adult life. Here, I have noticed aestivation among seven-spot ladybugs, in small clusters on the tops of thistles and wolf willow bushes, and among two-spot and polkadot ladybugs on the bark of trees.

Ladybugs have large eyes and a well-developed sense of sight. Ed Mondor and Jessie Warren found that adult Halloween ladybugs could associate yellow colour with prey, and they concluded that these beetles could find suitable plants visually in the field. The sense of smell is also apparently important for ladybugs. R.M. Hamilton and his colleagues in Utah looked at the sense of smell of convergent ladybugs, which are attracted to peach-potato aphids. They found that sensors on the antennae, near the tip, were responsible for the sense of smell. The ladybugs were attracted both to aphids on radish leaves, and to clean radish leaves, suggesting that a mixture of both scents may be attractive. These authors noted that before 1980 no one thought ladybugs could smell or see their food at any distance and were considered "blundering idiots" by most entomologists. Since then, both senses have been studied and confirmed, but not in a convincing way to everyone. The sense of smell seems to be present in convergent, seven-spot, Halloween, and eyed ladybugs (*Anatis ocellata*). The latter are attracted to the scent of pines, where they find pine aphids, their preferred prey. So a general pattern is emerging, and it seems likely that ladybugs rely quite a bit on their sense of smell.

Naturally, ladybugs also have many enemies, above and beyond their cannibalistic relatives. They are eaten by a variety of spiders and predaceous insects as well as by birds. However, according to Ivo Hodek and Alois Honěk "there is no tangible evidence of the impact of enemies on the population changes of coccinellids." Non-lethal attacks also occur, and it is common to find ladybugs with stain-like damage to their wing covers, a condition that has been interpreted as feeding damage by minute pirate bugs (family Anthocoridae) on ladybug pupae. These tiny sucking bugs merely wound the ladybug, while larger predatory hemipterans will kill them outright. It has also been suggested that although most songbirds come to recognize ladybugs as distasteful, flying insect eaters such as swallows and swifts may not recognize their warning colours in time to avoid swallowing them. I doubt this personally, since I can recognize a flying ladybug as distinct from other sorts of beetles myself, even at a considerable distance.

A cluster of two-spot ladybugs on a leaf in mid-autumn, using their combined warning colours to deter predators. They will move down into the leaf litter for the winter.

Playing dead and reflex bleeding are the primary defences that ladybugs have against their enemies. Since ladybugs feed on aphids, and aphids are often guarded by ants (who tend them for their honeydew in a tender sort of "mutualism" that is often likened to the relationship between dairy farmers and milk cows), ants are often a problem for ladybugs as well. This is probably why ladybugs have such short legs and antennae, and such smooth, rounded bodies. When threatened by an ant, they pull in their appendages and hunker down onto the substrate as tightly as they can. Finding nowhere to get hold of the beetle, the ant eventually gives up the struggle.

Ladybugs probably have more to fear from parasites and disease than from predators. They suffer from parastism by various tiny chalcidoid wasps, podapolipid parasitic mites, nematode round worms, sporozoans, fungi (especially *Beauveria bassiana*), and *Rickettsia* bacteria. Ladybugs are also parasitized by tachinid flies, phorid flies, and the braconid wasp *Dinocampus coccinellae*, but most of these seem to be more important ecologically in the Old World (although *D. coccinellae* can sometimes attack up to 20–30% of adult ladybugs in eastern North America). Of the ills mentioned above, only *Beauveria* seems really important in North America, since it can be, in humid areas, the main cause of death of overwintering ladybugs.

The *Rickettsia* bacteria are also significant sources of mortality, but in an odd way, since they kill only the males and are the subject of some very fascinating research, largely under the direction of Mike Majerus at Cambridge. For example, male-killing may actually be a valuable thing for the ladybugs, since they reduce the likelihood of a female mating with her brother, and the dead male eggs serve as food for the newly-hatched female larvae, reducing the likelihood of egg cannibalism. The phenomenon of male-killing bacteria seems restricted to aphid-eating ladybugs that lay their eggs in clusters—in other words, the subfamily Coccinellinae. The distribution of male-killing bacteria among North American ladybugs is still largely unknown.

Speaking of Mike Majerus, he is also the co-author, with John Sloggett, of an extraordinary paper in which they placed the origins of food and habitat preferences of ladybugs into a broad evolutionary context. In my opinion, this is one of those works of science that go far beyond providing just one more tidbit of information. Sloggett and Majerus suggest that new, specialized lineages may arise when generalized ladybugs feed temporarily but predictably in alternate habitats, on alternate prey. Some ladybug species have quite catholic tastes in habitat, while others are extreme specialist, and many lie somewhere in the middle. Those ladybugs with a predilection for life in extreme, alternative habitats may well remain in the alternate habitat early in the season, and breed there, and leave offspring that are likewise inclined to live in such places. Theoretically, natural selection could operate quite rapid in these instances, and this is the mechanism for "the origin of species" in ladybugs, as suggested by Sloggett and Majerus.

Both switching prey and moving between habitats, however, are likely to be costly to the beetles, since studies in these cases generally show a drop in ladybug fecundity. However, limited prey both early and late in the year may be an important motivating factor pushing ladybugs to search for novel prey types and, as a consequence, novel habitats. These alternate prey can include toxic aphids, low-quality aphids, aphids in galls with a defending soldier caste (believe it or not), and aphids tended by ants (some ladybugs are "myrmecophilous," living with ants, and this may explain the origin of this habit, seen in our own anthill ladybug).

In the entomological literature, scientists studying plant-feeding insects have suggested that these creatures quickly and predictably evolve to inhabit "enemy-free space," but this does not seem to be the case with predatory ladybugs. In the Sloggett–Majerus model, the important enemies are almost all parasites and parasitoids, since ladybugs are chemically protected from generalist predators. If we consider other ladybugs to be important predators, it naturally follows that a ladybug that lives in places without other ladybugs will indeed find itself in a sort of enemy-free space.

In general, zoologists have often suggested that geographic isolation is the thing that most predictably leads to the origin of new species, but this doesn't seem to be as important in ladybugs as it is in other groups (tiger beetles are a great example), since ladybugs are such good fliers. Here in western Canada, for example, Joe Belicek concluded that the Rocky Mountains are not a barrier to the dispersal of ladybugs into Alberta from the west. Species that occur on only one side of the continental divide do so because they find appropriate habitat there, not because they can't reach the other side.

Aphid-eating ladybugs are better at dispersing than scale-insect eaters, and they typically change habitats throughout the season, sometimes predictably moving from one tree or vegetation type to another. This is when they sometimes wind up in lakeshore "wash-ups," a subject that we will revisit in the next chapter. Oddly, and interestingly, Sloggett and Majerus point out that in the few studies that have investigated it, there is little ecological niche overlap among ladybugs. Instead of feeding and competing in the same place, they divide up the habitat according to such things as height above the ground, degree of sunlight exposure, and the like.

The Sloggett–Majerus model is appealing and a great bit of synthesis in an otherwise largely piecemeal field of study. I think it would be especially interesting to look at it in the context of introduced species, and Edward Evans' notion of "habitat compression," in which native species "retreat" to their "ancestral habitats" in the face of superior competitors from Europe or Asia. I suspect that, rather than retreating to ancestral habitats, species such as the transverse ladybug are probably quickly evolving to use new refuges from the seven-spot. We may well be seeing evolution in action here, but I'll come back to this theme in Chapter 4.

Edgar H. Strickland, Alberta entomology pioneer, examining a yucca plant on the open prairie, no doubt alert for rare ladybugs at the time.

3
Ladybug Study in Alberta

The scientific study of ladybugs in Alberta began with beetle collectors. Frederick S. Carr was one of the most important of these, and his story appears in my book on tiger beetles. In my own database of ladybugs records, the oldest entry is for a nine-spot ladybug, collected by F.S. Carr in Medicine Hat on May 3, 1903. In fact, the early ladybug collectors in Alberta are all men who feature in the tiger beetle and damselfly stories as well. J.B. Wallis, from Manitoba, was a school inspector and amateur entomologist, like F.S. Carr. Colonel Edgar H. Strickland, on the other hand, was the founding chair of the Department of Entomology at the University of Alberta. None of these men did much to analyze our ladybug fauna, but their specimens are all we have to reconstruct the ladybug situation in Alberta a century ago. I have found the Carr and Strickland specimens to be especially interesting in this regard.

By the 1950s, the number of beetle collectors had increased by quite a bit, and Alberta ladybug specimens were making their way to such luminaries as Theodosius Dobzhansky. Dobzhansky was not only a ladybug taxonomist, he was also an important geneticist, and one of the key figures in what we now know as the neo-Darwinian revolution, in the mid-1900s. During this time, Charles Darwin's ideas on natural selection were reexamined in the light of genetics, paleontology, ecology, and other aspects of biological science. The synthesis that resulted has been at the core of biological thought ever since, and it is particularly gratifying to know that a ladybug scientist was one of the prime contributors (although in fairness much of his research was done on fruit flies as well). Dobzhansky was born in the Ukraine (then part of Russia), and came to the United States in 1927, where he did most of his work at Columbia University and the California Institute of Technology. He was a deeply humane person, as are most evolutionary biologists (despite what some anti-evolutionists might think), and his 1967 book "The Biology of Ultimate Concern" is proof of that. He was one of many biologists who argued that we should set aside the conflict between science and religion and get on with both. Dobzhansky is probably best known for the quote "Nothing in biology makes sense except in the light of evolution;" and for the fact that Stephen J. Gould of Harvard referred to him as "the greatest evolutionary geneticist of our times."

The first work specifically on the ladybugs of Alberta was published a year after Dobzhansky's death, in 1976. Joseph Belicek's study was based on his Masters degree in entomology, completed under the supervision of George Ball (whose name also appears often in my tiger beetle book). Joe studied the ladybugs of western Canada and Alaska, and speculated on the origins of the fauna as well as surveying the known records at the time. He recorded 64 species from Alberta.

In Belicek's opinion, most of our ladybug species spent the last ice age south of the glaciers in the United States, but some may have survived in the unglaciated Beringian region in Alaska or the ice-free area around Mountain Park in the Rocky Mountains. After all, even in an ice age, there are summers and winters, and the summers may have been warm enough for ladybugs to survive. Joe also concluded that the Rocky Mountains were not a barrier to the dispersal of ladybugs into Alberta from the west. As well, Joe described two new species from Alberta, the Hercules and Jasper ladybugs. Joe's paper was an inspiration to a number of beetle collectors and provided the foundation on which the subsequent study of the ladybugs of Alberta has been built.

The next person to look at our ladybugs in detail was Robert D. Gordon of the Smithsonian Institution in Washington, D.C. Bob published many papers on the ladybugs of North America, but his most important works were a huge revision of the subfamily Scymninae and a marvellous monograph on the North American ladybugs in general that took up an entire issue of the Journal of the New York Entomological Society. When it appeared, I was a graduate student, and I remember fellow grad student Bob Anderson coming down the hall to tell me to immediately join the New York Entomological Society, since that was the only way to get Bob Gordon's coccinellid monograph. So I did, and I'm glad—the book (it really is a book, not merely a paper in a journal) is now almost impossible to obtain now, even with on-line searches of used book dealer's lists.

Whenever Bob Gordon's name comes up among entomologists, someone invariably says something like "his coccinellid monograph is magnificent, but did you know that his true passion is for scarab beetles?" It's true, apparently, and Bob has made great contributions to the study of scarab beetles (as well as beetles on sand dunes) outside his work on ladybugs. However, the Smithsonian wanted someone to work on ladybugs, primarily because they are important to agriculture. Bob spent a great deal of his time keeping track of the various ladybugs that had been introduced to North America for aphid and scale insect control. One of his most useful papers (on identifying introduced species) was co-written with Natalia Vandenberg, the woman who took Bob's place when he retired. "Nat" is a true lover of ladybugs and a very fine scientist as well. Unlike most beetle systematists, she does not go out and collect speci-

Top: A Hudsonian ladybug takes to the skies, risking falling into a lake and washing up on shore later in the day, when beetle hunters survey the "wash-up." Bottom left: Pioneer beetle collector F. S. Carr. Bottom right: Joe Belicek, who reviewed the ladybugs of Alberta in 1976.

Top left: Patsy Drummond, disappointed at the wash-up at Lesser Slave Lake after widespread forest fires in 2001. Top right: The author and Shelly Ryan digging up overwintering ladybugs in a pine forest near Opal. Bottom: Mira Snyder searches for ladybugs among rocks along the shore of Waterton Lake.

mens, since she has more than enough work studying the specimens already in the Smithsonian collections, and she really doesn't like the idea of killing ladybugs. She doesn't object to collecting; she just doesn't do it herself. I like Nat. She keeps singing crickets in traditional oriental cages and is involved in folk music—a well-rounded human being.

Next on the list of ladybug events in our province was the somewhat ill-fated national ladybug survey, initiated by the Canadian Nature Federation from 1995 to 2000. The objective of the survey was to monitor the effects of introduced species, but the inadequate identification guide the survey was based on made most of the data collected (some 32,579 records) difficult if not impossible to interpret. I'll tell you more about this story when we come back to the subject of native and non-native species.

In Calgary, Ed Mondor did some interesting studies of ladybugs and their response to the alarm chemicals given off by aphids, but his time there was limited, on a post-doctoral appointment, and he has gone on to work in the Hawai'ian Islands now.

Over the past decade, I have been fortunate to work with a number of talented undergraduate students with an interest in ladybugs. Much of what I know I learned from their studies, and it is no exaggeration to say that they collectively laid the groundwork for this book. In 1996, Shelley Ryan was a student at the Northern Alberta Institute of Technology, and she decided to do her final project in the biological sciences program on the seven-spot ladybug. She came to me for help, at the suggestion of her instructor Robin Leech, and we set up some simple experiments to see how well the seven-spot handled Alberta winters. Surprisingly, over 97% of the beetles survived the winter (compared to about 50% in Europe), and in experimental cages the temperature did not drop below –9 °C, even when the air temperature dipped below –30 °C. We suggested that seven-spots benefit from this in North America, and that here they are not as susceptible to overwintering mortality from *Beauveria* fungus infection. As well, Shelley looked at the fecundity of captive transverse and seven-spot ladybugs and found them to be roughly equivalent. Since then, Shelley has informally repeated some of her work and found that the initial studies benefited from unusually deep snow—survival in subsequent winters was not as high. Still, Shelley keeps up an active interest in ladybugs and has contributed many records to our database.

My first undergraduate ladybug student was Wendy Harrison (now Wendy Wheatley). I teach in the Department of Renewable Resources at the University of Alberta, as a Sessional Lecturer. In other words, I am at the bottom of the academic totem pole in some ways; however, I am also spared the burden of administrative work, and I am free to pursue projects such as this book, rather than focusing on obtaining grant money and publishing research papers. I can't supervise graduate students, but I do supervise undergrads doing independent research. Wendy told me of her interest in insects, so I suggested a project to her, in which she collected historical records for Alberta ladybugs and analyzed these to see if she could detect evidence that the introduced seven-spot ladybug had caused the decline of native species. This project was motivated by a review of Shelley Ryan's and my paper, by Cedric Gillott from Saskatchewan (again, there is more about him in the damselfly book). Cedric had quite rightly asked us if perhaps the seven-spot had become abundant without affecting other species and if the so-called declining species might only be relatively (but not absolutely) less common.

Wendy collected all of the specimen and observational data she could find and then calculated the relative proportion of the fauna (not including the seven-spot) each year after the seven-spot arrived in Alberta for a ten-year period. Of course, we only expected a rough indication of abundance trends, since the specimens had been collected haphazardly, by other people, with no particular point in mind. Acknowledging this, Wendy performed linear regression analyses on each of the species' data with a large enough sample size. Linear regression is a technique for determining whether there is a trend in your data (and if you've ever gathered data, you will know how fuzzy these trends can be!), how strong the trend is, and how steeply things are either increasing or decreasing over time. Wendy's study showed that at least some species (the transverse and the convergent) appeared to have decreased in abundance among the native species as a whole, while two (the two-spot and the polkadot) actually seemed to increase, at least in relative terms (Table 1). These latter two species may not have changed at all in their absolute abundance, but they would have appeared to become more common if the rest of the native fauna had declined relative to them. Interestingly, four native species (the thirteen-spot, parenthesis, painted, and the wee-tiny ladybug) did not change in relative abundance, and when the overall fauna was analyzed, the seven-spot did not show any trend either, suggesting that it became the most abundant species very quickly. I'll come back to these results later; however, at the time Wendy's study was both perplexing and tantalizing, and she did a fine job of presenting it to the annual meeting of the Entomological Society of Alberta at their annual meeting in the fall of 2000. Since the relevant statistical measures for these regression analyses were not published in the abstract of that talk, here they are for posterity. Note that a high r^2 indicates a clear trend without much scatter around the trend line, while a P lower than 0.05 indicates a "real" trend, not likely due to chance alone.

Another student, Patsy Drummond, worked as a park interpreter at Lesser Slave Lake Provincial Park, and in 2001, she made weekly searches of the beach at Lily Creek. She managed to count 1778 ladybugs (representing 14 species) early in the season, but forest fires seemed to reduce ladybug activity to almost nil past June 18. Patsy therefore focused on other things, including a complete reworking of our provincial ladybug database, the preparation of new range maps for all of our larger ladybug species (which were the basis for the maps

TABLE 1 Linear regression statistics for change in relative abundance (percentage of total larger ladybug fauna excluding *Coccinella septempunctata*) from 1989 to 1999 of all ladybug species for which sufficient data was available.

Species	Slope	$F_{1,7}$	P	r^2	Error
Decreasing abundance					
Transverse	−0.03	5.36	0.05	0.43	0.11
Convergent	−0.05	16.02	0.01	0.70	0.13
Increasing abundance					
Two-spot	0.05	11.52	0.01	0.62	0.15
Polkadot	0.01	17.56	<0.01	0.71	0.02
No change					
Thirteen-spot	0.01	0.37	0.56	0.05	0.12
Parenthesis	0.00	0.03	0.87	0.00	0.10
Seven-spot*	−0.02	0.76	0.41	0.10	0.23
Painted	0.00	0.06	0.81	0.01	0.16
Wee-tiny	0.01	0.50	0.50	0.07	0.08

The analysis for C. septempunctata *itself included all larger ladybugs data for the period.*

in this book) and a general assessment of the conservation needs of Alberta's ladybug fauna. I was especially pleased that Wayne Nordstrom at the Alberta Natural Heritage Information Centre gave Patsy space in which to work and made her feel welcome and useful at this government office.

Then, in 2002, Mira Snyder completed a survey of the ladybugs at Waterton Lakes National Park. Mira worked there during the summer, and we obtained a permit for the research. The objective was to see if the Halloween ladybug had arrived yet, since it would be expected first in the southwest corner of the province. Mira did not find the Halloween, but she did collect great data from the beaches of Waterton Lake and from sweeping vegetation at various locations in the park. Mira found 14 species (out of 27 historically recorded for the park) even though she only encountered 153 ladybugs over the season. Like Patsy, Mira was plagued by bad weather, and in Mira's case it was early summer snowstorms that likely reduced the numbers of ladybugs.

Mira also analyzed wash-up data from around the province and was surprised to find that there was no increase in overall ecological diversity from south to north, even though the northernmost wash-up samples, from Lake Athabasca, had relatively few seven-spots in them (only 19%). It is worth mentioning here that ecologists do not simply count species when they assess diversity. They have also developed indices of diversity that take into account the relative abundance of all species in the fauna. For most people, including many trained biologists, it is difficult to believe that a fauna containing one or more super-abundant newcomers can be more diverse than the fauna that came before it. However, these people, and their intuitive notion of how things work, are simply wrong. Most of the time, introduced species increase biodiversity (no matter how you measure it), and that is the simple ecological truth of the matter, even if it goes against the invasion biologists' cry that such newcomers are by definition a threat to biodiversity.

Wash-ups have proven to be some of our best sources of information on ladybug numbers. A wash-up happens when beetles in flight fall into a lake and wash up on the shore. In general, this happens most readily on warm afternoons, with a storm brewing at the end of the day. The beetles (ladybugs, as well as a variety of other beetles, as well as some other insects) are usually alive when they reach the shore, and if the temperature is still warm and the sun is shining, they quickly take flight again. But more often, they accumulate in a sort of windrow, and can be counted easily.

We have wanted to assume that the composition of ladybugs in a wash-up is a good measure of the composition of the fauna in the surrounding area, but although there is little reason to think otherwise, there is also little evidence that this is true. In the US Midwest, Richard Lee found that some ladybugs remain at the wash-up for two to three weeks and suggested that migration to and from overwintering sites might account for wash-ups. I suspect this is not the case here, because wash-ups occur all through the season, not just in spring and fall. As well, there are other questions. Are all species in an area equally likely to wind up in the lake? Are all species in the lake equally likely to wind up on shore? How long do individuals live once they hit the water? How far off shore do the ladybugs come from? Do winds force them down into the water, or do they just fall in? Is there any reason to think that they fall in the water on purpose? Are all wash-ups afternoon phenomena? Is ladybug survival/fecundity affected by being in a wash-up? Are ladybugs adapted to breathe while they float? Are all wash-ups the result of dispersal for food, or for overwintering sites as well? The list goes on. Also, wash-ups are not a universal occurrence. In many places, they are almost unheard of. Thus, we are probably in a good position here to study this phenomenon.

My own interest in ladybugs began as a child and as a beetle collector. Then, when we were preparing the pilot episode of my television series "Acorn, the Nature Nut," I chose ladybugs as the subject, because there are ladybugs almost everywhere in the television-watching world. That was when I realized that the transverse ladybug of my childhood had been replaced by the seven-spot ladybug from Europe. I began keeping track of the ladybugs I encountered and collecting the literature on ladybug science as well. As my students worked on their research projects, I compiled notes for this book.

Mike Majerus gives a spruce tree a firm but gentle beating, hoping to dislodge some rare beetles.

I should also give credit to Mike Majerus here. When Wendy Harrison began her studies with me, I instructed her to read Mike's New Naturalist volume, "Ladybirds" cover-to-cover, and we discussed each chapter together on a weekly basis. At the end of this process, I wrote to Mike, who is the Reader in Evolution at the Department of Genetics at Cambridge University in the United Kingdom. I told him that we had enjoyed the book, despite its tight focus on the British fauna and, therefore, its limited applicability to the situation in Alberta. Mike wrote back and told me that he was working on a new book with a broader geographic focus and was coming to North America in the summer of 2000 to familiarize himself with the North American fauna. I suggested that he "stop by while he was here," and so he did. Mike and his family landed in California, drove to Alberta, stayed with my family for 11 days, and then drove to Miami, Florida. During our visit, I learned more from Mike about ladybugs and how to study them in the field than I could have absorbed in years of reading. He has become a good friend, and I was happy to be able to invite him back, with his wife Tina in 2005, to serve as the Plenary Lecturer for the joint meeting of the Entomological Societies of Alberta and Canada. Mike has focused his career on both ladybugs (which he calls ladybirds, in keeping with British usage) and moths and is probably best known for defending the classic work on the peppered moth and industrial melanism, a textbook example of evolution in action that has recently come under critical scrutiny and therefore needs a champion and some fresh data.

The margins of this foothills stream are home to some of the wetland-dwellng ladybugs, including the rare waterside ladybug.

4

Introduced Ladybugs and Conservation

Without a doubt, the biggest news in Alberta's ladybug history has been the recent arrival of non-native species and their effects on the native fauna of our province. The *Edmonton Journal* reported the story, on Sunday, July 9, 1998 (page A5):

Problems "Spotted" in Ladybugs' Rivalry
Ross Henderson, Journal Staff Writer

She may be a ladybug, but she ain't no lady.
In fact, in the tiny world of ladybugs, a battle for insect supremacy is being waged between two of Canada's predominant species.
No kidding. Right below your feet.
The spoils of the war? The 473 other species.
The prospect of one species taking over in Canada has convinced the Canadian Nature Federation in Ottawa to launch a study that's expected to continue for several years.
The non-profit federation is using an Internet page and sending out thousands of kits to school kids, libraries, and other groups to help spot ladybug species.
Marc Johnson, a spokesperson for the federation in Ottawa, says two species are battling it out over pests and aphids, their favourite food. Both the competing ladybugs were foreign to Canada, and as a result, lack our peacekeeping skills....

Further on in the article, Henderson writes "As the nineteen-spotted, southern and seven-spot ladybugs take over turf, the diversity of species—which assures a biological safety net—is lessened and reduces the delicate balance of the ecosystem." The article ends with the chilling observation that

The seven-spot ladybug: an unfair, alien, super-competitive invader, or a hard-working new immigrant that deserves its new place as the most common member of the Alberta ladybug fauna?

"the nine-spotted ladybug was, until recently, the most prevalent species in Ontario. It can't be found now."

Like most naturalists, I was convinced that this was a major ecological disaster. I would not have gone so far as to suggest that our "biological safety net" was in danger (I deeply dislike environmentalist hyperbole, and as much as I love ladybugs, I'm sure we could survive without them; many other creatures also eat aphids, scale insects, and spider mites), but I was upset nonetheless. During the mid-1990s, as I searched in vain for the transverse ladybug, the most common species of my childhood, I came to the conclusion that this was the most profound ecological change I had ever witnessed in a lifetime of insect studies in Alberta. So I began spreading the bad news, on my television show, in public talks, and in university lectures. I also started submitting my own observations to the Canadian Nature Federation Ladybug Survey.

The ladybug survey was circulated to schools, naturalists, and garden centres. It was immensely popular, with over 45,000 survey form requests received in a single year. As the Federation put it, "the purpose of the lady beetle survey is to determine the geographical range of native and non-native lady beetle species in Canada. Once we have determined these ranges, we can find out if alien lady beetle species inhabit the same areas as our native species." However, it was already apparent to naturalists across the country that this was clearly the case. The survey's real purpose, in my opinion, was to affirm the obvious (not that this is unusual in science) using citizen-science data and to spread awareness of the dangers of introduced species. The Seeing Spots Newsletter wrote: "A major concern among survey participants is the presence of alien lady beetle species in Canada," and although the Federation never really claimed it could do anything to get rid of the so-called "aliens," this was clearly the implied goal for many of the people I spoke to. It was not uncommon to watch naturalists crush seven-spot ladybugs as soon as they recognize them as unwelcome invaders. I was an active participant in the survey, and I was as concerned as anyone about the "invasion" at the time, but I didn't kill seven-spots.

"Invasion biology" is a relatively recent development, and the scientists who work in this field are primarily concerned with the negative impacts that introduced species have on agriculture, forestry, or other industries. Invasion biologists see their discipline as originating with the publication of Charles Elton's "Ecology of Invasions by Animals and Plants" in 1958. According to my friend Matt Chew, a Ph.D. student at Arizona State University, there were numerous other biologists who laid the groundwork before Elton, but it was Elton who had the highest profile and who established the use of military and political metaphors to describe the arrival and spread of new species in terms such as "alien invasion," "take-overs" and "battles." Go back and read the newspaper clipping at the head of this chapter and notice how the language is emotive and military, not objective and scientific. Elton, along with his good friend the conservation pioneer Aldo Leopold, also invoked aesthetics as a primary justification for anti-alien sentiment, especially in the field of conservation. In other words, he thought that in any given place "native" organisms and ecosystems were intrinsically more beautiful than non-natives and that this should be apparent to everyone. The concept of nativeness, by the way, is a tough one to meaningfully define biologically (after all, living things move around, and always have), and this is only recently becoming apparent, again largely thanks to Matt Chew and a small group of equally courageous scholars.

Invasion biologists have declared that invasive species are a major threat to the biological integrity of the planet, and these people have recently received a great deal of attention and money. It is important, however, to recognize that they are primarily applied biologists, working on problems of economic importance (often with the blessings of both government and the pesticide industry). Biologists who are more interested in the workings of nature than in the affairs of people have been less alarmist in their interpretations, and many have accused the invasion biologists of showing strong cultural biases. These include a tendency toward racism (in the form of "nativism"—a prejudice against non-native species), xenophobia (the fear of things that are new or unfamiliar), nationalism, and stabilism (the belief that things should forever stay the way the Creator, or the "balance of nature" intended them). However, back in the late 1990s, the debate about the motives and biases of invasion biology had not yet surfaced, and it certainly hadn't filtered down to the ladybug crowd.

As the ladybug survey came to an end, it was obvious that it had not been entirely successful. The website had some distribution maps posted for a while, but the data were bizarre and the maps did not remain on the site for long. It was clear to those of us who really knew ladybugs that misidentifications were common, and that the colour pamphlet that came with the survey was over-simplified to the point where misidentifications were almost a certainty. As soon as I realized that the pale, prairie morph of the two-spot ladybug could be mistaken for the Halloween, I gave up on at least this aspect of the project, and it is now clear that the organizers gave up as well, albeit still saving face, and making the best of a messy data set.

I did, however, print off the "results" section of their website on August 11, 2001, at http://cnf.ca/beetle/spotted_results.html. They received a total of 32,579 reports, 60% of which were either seven-spots or Halloween. They found the seven-spot across Canada and apparently "moving northward" (something the survey could not have detected, in my opinion). They also found the Halloween to be "well established across the country," which it is not, and the fourteen-spot ladybug (*Propylea quatuordecimpunctata*) to be "successful in establishing itself across the country," despite its total absence from Alberta and perhaps other parts of the west. The American eyespot ladybug was "historically found across Canada, but a scant 220 sightings indicate a sparse presence in the east, with few sightings in the west." This didn't surprise me since here in the west the species has always been uncommon (and 220 sightings have to count for something). The report finished with the notion that alien species seem to be causing the decline of natives but that the survey can't prove it for sure. No summary publication appeared, and no great truths were revealed. The main effect, then, was to spread fears about "alien invaders" among the naturalist community in Canada. As I write this, there is a new version of the "results" on the web, acknowledging the weaknesses in the data but perpetuating the anti-invader theme.

So what really happened, and what is happening now? Let's first review what we know from data in Alberta. Wendy Harrison and I gathered all the records we could from naturalists and collections and determined that transverse and convergent ladybugs declined after the arrival of the seven-spot. Four other species showed no such trend, and for the rest of the fauna, we don't have adequate data to say one way or the other. I have already mentioned this in the last chapter, but it bears repeating here.

Now let's go back and start at the beginning. R.L. Jacques found the first North American seven-spots in 1973 in New Jersey. Seven years later, Richard Hoebeke and A.G. Wheeler Jr. mapped the distribution of the seven-spot, and in 1986 Paul Schaefer, Richard Dysart, and Harold Specht looked again at its distribution. By that point, it was still mainly an eastern species, nowhere

Two well-known newcomers to Alberta, the European skipper and the house sparrow, neither of which has "taken over" the ecosystems of the province.

Top: Just the right conditions for surveying ladybugs at wash-up—a warm afternoon with a storm brewing in the west. Bottom: The nine-spot ladybug, a species now extirpated from much of eastern North America.

near Alberta. The latter paper added that at a mass wash-up in Delaware, some people were bitten by the new ladybug, thereby suggesting that it was both ecologically and behaviourally "aggressive." Europeans knew that the seven-spot might occasionally bite, but everyone realized this was trivial.

It was clear that the seven-spot was headed our way, and in Nova Scotia, David McCorquodale estimated that the four species of "adventive" ladybugs (I like that neutral term, as well as the term "newcomer") in that province had spread at rates between 30 and 400 km/year. Bob Gordon and Nat Vandenberg pointed out, however, that the seven-spot got to the west at least partly with the help of people, since it was being released to control the Russian wheat aphid (*Diuraphis noxia*) in the west, a newly established pest of wheat that arrived in the late 1980s. The ladybug doesn't control the aphid, by the way, but that disappointment seems to have been forgotten in the ensuing kafuffle over its so-called rise to dominance.

In West Virginia, M.W. Brown and S.S. Miller were monitoring their own changing ladybugs. They started with 25 species in 1983, in their study area (the total ladybug fauna of West Virginia is much greater than that). Then, from 1989 to 1994, the seven-spot arrived and "dominated the fauna." In other words, it became the most common species. In 1995, the Halloween ladybug arrived in West Virginia, and "displaced" the seven-spot as the most common species. The seven-spot comprised 100% of the fauna in 1992 ("all but one" record), then dropped to around 90% in 1993, before the arrival of the Halloween in 1994, when it dropped again to about 80%. By 1996, the seven-spot made up about 10% of the fauna, and the rest were Halloweens. Clearly, there were few native species left to count. Brown and Miller did find a bit of good news in all of this—the Halloween ladybug did a better job of controlling the green citrus aphid (*Aphis spiraecola*) than did the seven-spot, or the native species that preceded it.

It is easy to bemoan the loss of the native fauna, and the short sightedness of entomologists who cared only about aphid control, but there are also reasons to remain skeptical of the notion that all is lost. In their encyclopedic work on ladybug ecology, Hodek and Honěk state, "coccinelid communities contain a few dominant species. Usually two to four species represent more than 90% of the individuals." They also point out that "there is no record of a global extinction of a coccinellid species since the beginning of scientific interest in the family." Is it possible, then, that what we are witnessing is the replacement of one "dominant" species by another and not much ecological change among the less abundant species in the fauna? If so, we need studies that survey ladybugs in all of their different habitats, carefully searching for each and every species, and not just sweeping a net through monotonous agricultural crops and fruit orchards.

In Maine, Andrei Alyokhin and Gary Sewell looked at 31 years of ladybug data, collected from 1971 to 2001, but entirely in potato patches. At the beginning of the study, the most common species were the native transverse and thirteen-spot ladybugs, although there was tremendous year-to-year variation in the relative abundance of both. In 1980, the seven-spot arrived and became the most abundant species in the study. In 1995 and 1996, respectively, the Halloween and the fourteen-spot arrived, and they became common as well. The effect of their arrival was not only to decrease the numbers of transverse and thirteen-spot ladybugs but also to increase the overall diversity, measured by diversity indices (ecological measures that incorporate not only the number of species but also the relative abundance of each). Only when the Halloween ladybug arrived did aphids detectably decline in abundance, and of course that is what the potato farmers were interested in. Transverse and thirteen-spot ladybugs persisted in the potatoes, but the authors suggested that the arrival of the new species caused the old species to retreat to ancestral habitats in a process called "habitat compression," citing Edward Evans, and a paper he published in 2000.

Just last year another Mainer, a graduate student of Andrei Alyokhin named Christine Finlayson, investigated the habitat compression hypothesis in habitats other than potato fields. With the help of Andrei Alyokhin and Kristine Landry (an undergraduate field assistant), she discovered that the non-native species were clearly dominant in all habitats, including the "ancestral" habitats such as forests and riparian areas, and the native species showed no tendency to do best in so-called natural places. Thus, the habitat compression hypothesis was not supported by this study. In a letter to me on February 7, 2006, Christine wrote, "lady beetle diversity was shown to increase as the number of non-native species increased. However, native lady beetles were found in very low numbers in all of the habitats surveyed. *H. tredecimpunctata* [the thirteen-spot] and *C. transversoguttata* [the transverse], the two native species that were once dominant here, made up only 1.09% and 0.07% of the total aphidophagous lady beetles collected, respectively." I think this work is a major step in the right direction and that the next logical step should be to take the naturalists' approach and search for the native species one at a time, exploring a wide variety of microhabitats rather than broadly surveying hypothetical ancestral habitats.

In Manitoba, the seven-spot arrived in 1988, and two biologists, Glen Wylie and Frank Matheson were keen to document its arrival in crop fields. They therefore began systematically sampling the crops with sweep nets. Earlier, in 1978, Bill Turnock and Russ Mead (of the Delta Marsh field station) began recording wash-up aggregations on Lake Manitoba at Delta Bay (when the convergent ladybug made up 96% of the wash-up) Turnock and Mead were initially interested in the wash-up for it own sake and noted that wash-up happens "whenever a northerly wind follows a warm, windy day." They sampled the following morning, and recorded their highest counts in spring and fall during ladybug movements to and from over-wintering sites, in keeping with what we have seen here in Alberta. Ian Wise has also been continuing the crop sweep sampling begun by Wylie and Matheson for Agriculture Canada. Encouragingly, the results of the two surveys are very similar, suggesting that wash-ups are indeed a good way to survey local ladybugs. The seven-spot rose to 67% of the sample in 1992, then dropped to 4% in 1994 (a cold summer when the thirteen-spot made up 95% of the sample), and now varies from 20% to 35%. Their conclusions? The seven-spot ladybug has not affected the thirteen-spot but probably has affected the convergent and the transverse (as well as parenthesis and three-banded ladybugs). Or perhaps the thirteen-spot has somehow indirectly benefited from the change in fauna. Either way, what they recorded were changes in species abundance, not local extinctions.

There are other studies as well. In South Dakota, Norman Elliot and his colleagues found that, when seven-spot ladybugs arrived the abundance of transverse ladybugs in alfalfa fields declined to between 3% and 5% of its original level but that overall ladybug numbers did not increase, suggesting that the alfalfa environment could only support a certain number of ladybugs. In the northeastern United States, Donna Ellis and her colleagues suggested that the spread of non-native ladybugs had reduced the abundance of nine-spot and convergent ladybugs. In Tennessee, in non-agricultural "natural" habitats, a group of researchers from the University of Tennessee failed to find the convergent ladybug and thought that perhaps the seven-spot and Halloween had resulted in its absence.

In contrast, Georgia researchers Leonard Wells and Robert McPherson wrote that "although we have no record of *H. convergens* population levels in Georgia tobacco before invasion by *C. septempunctata* and *H. axyridris* [the seven-spot and Halloween ladybugs], these exotic coccinellids do not as yet appear to have had any adverse effects on *H. convergens* populations in Georgia tobacco." These researchers did find that the seven-spots appeared to leave the tobacco once large numbers of Halloween ladybugs had arrived, and cited a similar study that showed the same thing on pecans, by Louis Tedders and Paul Schaefer, in 1994. Tedders and Schaefer chronicled the establishment of the Halloween ladybug in the American southeast, and commented that the pecan growers were happy with the way it controlled two species of aphids.

The cause for the success and spread of introduced ladybugs is not at all clear. Certainly, both seven-spot and Halloween ladybugs can lay huge numbers of eggs, but this fact alone does not explain why they have done well in North America—after all, they coexist with hundreds of other ladybug species in Europe and Asia. Most explanations for the success of the newcomers are based on assumptions about competition. For example, Edward Evans found that the seven-spot is more variable in body size than related native species, and considered this to be evidence that it was more of an ecological generalist, adaptable and a good competitor.

Competition is often assumed to be at the heart of "alien invader" success in general, and at least one author, David Theodoropoulos, has suggested that the same sorts of sentiments that make people angry at new immigrants "taking our jobs" motivate the thinking of some ecologists when confronted with "new immigrant" species. Theodoropoulos calls himself an "invasion skeptic" and although his book "Invasion Biology: Critique of a Pseudoscience" is unlike your average scholarly book and more of a rant against the ecological establishment, he makes some very interesting points, especially with respect to misplaced assumptions about invasion biologists' ideas regarding so-called unfair competition. In that context, it is interesting that Edward Evans found that similar-sized larvae of different ladybug species were roughly equivalent in terms of their competitive ability, and in their effect on aphid numbers (working in the lab with convergent, sinuate, thirteen-spot, and seven-spot larvae). I for one am skeptical that competition has much to do with the success of our new ladybugs, and I readily admit that the more I work with ladybug studies the more I see myself as a bit of an "invasion skeptic" as well.

Two studies by John Obrycki and his colleagues (both published in 1998) illustrate the complexity of the issue, and how little we actually know. Both studies involved seven-spot and twelve-spot (*Coleomegilla maculata*) ladybug larvae. In one study, it was suggested that when there are more aphids, the larger seven-spot ladybug larvae have the advantage because they can bully the smaller twelve-spot larvae (bullying is technically called "interference competition"), but when there are few aphids the smaller twelve-spot larvae have the advantage because they need less food. In other words, they can coexist. In the second study, it seemed that the seven-spot wins out at low aphid densities either because it out-competes the twelve-spot or simply eats them.

In my opinion, the success of introduced ladybugs probably has to do with the general nature of ladybug guilds and a variety of quite simple factors that are nonetheless difficult to detect or measure. Yes, the new ladybugs reproduce well, they live in a variety of habitats, and they do well in the sorts of habitats that people produce, including agricultural crops, orchards, and shrubs. That is why they were introduced in the first place. We also know that ladybugs eat each other, and this is especially true of larvae eating eggs and of larvae eating other larvae. We also know that it is typical and normal for a few species to "dominate" the rest in any given environment. So, is there really any mystery why the seven-spot and the Halloween are common? The real mystery, in my opinion, is why there are so many species of ladybugs in the first place and how they coexist!

One piece that is completely missing from this puzzle is the state of the ladybug fauna before we began the scientific study of the insects of North America. In Alberta, our record goes back just a hair past 100 years, and there is no doubt that most of what we know about what was "normal" before the arrival of the seven-spot is based on ladybugs already inhabiting such man-

The transverse ladybug: once the most common species in Alberta and now a minor but persistent member of the fauna.

made environments as crops, gardens, parks, and the like. The species that were formerly common, such as the transverse ladybug, were also common in man-made habitats, and it is entirely possible that such species built up their numbers in disturbed areas before flying into natural areas only to be collected by entomologists and counted as part of what was assumed to be the normal fauna. After all, we can find adult seven-spots in places where we cannot find their larvae, such as above timberline in the Rockies. Perhaps other species have been spending time in marginal habitats all along. Other species, such as the thirteen-spot, tell a different story. In natural habitats, it is pretty clear that, both here and in Europe, the thirteen-spot is a species of marshes and grassy wetlands. Yet it also does well in lawns, gardens, and some sorts of cropland—places that did not exist here 300 years ago. Does that make it an abundant native species or a former habitat specialist that has subsequently "invaded" man-made environments? It seems to me that, for the most part, we cannot answer these questions, and the relevant evidence is forever lost in times past. The importance of disturbed or "man-made" habitats, however, should not be underestimated.

Returning to the transverse ladybug, I find this our most intriguing species. It was once super-abundant in all manner of habitats in Alberta. Then, with the arrival of the seven-spot, it almost disappeared. Wendy Harrison and I backed this up with data, but any good naturalist could tell what was happening at the time. During the late 1990s, we would find a transverse here or there, but always alone, and always in the company of huge numbers of seven-spots. Then, during Mike Majerus' visit, a different pattern began to emerge. One afternoon, while we were preparing to host a party in Mike's honour, I dropped Mike off south of Edmonton at the University of Alberta's Sandy Mactaggart Nature Sanctuary (at a spot that is now part of the Anthony Henday freeway), and the rest of Mike's family at the West Edmonton Mall. When I came back to get Mike, he told me that he had spent many hours searching the "good" habitat, only to realize that the ladybugs were instead living on the dry pediment slope at the base of the reclaimed coal mine hill, among scraggly grasses (a spot that is now underneath the new ring road around west Edmonton). There, he found only 2 seven-spots, but 21 transverse, along with about 20 larvae and 7 pupae (along with a good number of parenthesis ladybugs and their larvae). We were both puzzled—what were they eating, and why were they living in this harsh, hot, dry microenvironment.

I have since found more transverse ladybugs in this sort of habitat, which is typical of Alberta badlands where the Cretaceous sandstones weather in such a way as to create nearly horizontal pediments at the bases of badland hills. This habitat, by the way, is much more common than the ladybugs. But the story doesn't end there. With Mike and his family, we also found transverse ladybugs (27 of them, along with 30 seven-spots) on the crest of partially vegetated open sand dunes northwest of Opal. Shelley Ryan and I had noticed before that this habitat sometimes produced transverse ladybugs, and since then I have found that it is probably the best place to look for them. Prairie sand dunes (such as the Great Sand Hills of Saskatchewan) are also good places to search for transverse ladybugs and nine-spot ladybugs as well. The situation at Opal is confusing, however. On one occasion we found three tamarack ladybugs, and no transverse, and on another occasion the dune ridge was home to plenty of seven-spots but neither of the other two species.

If we apply Edward Evan's idea of habitat compression to this species, it would seem obvious that the transverse is retreating to its ancestral habitats (ancestral in the sense of where it lived before the advent of human development and agriculture) in the face of the seven-spotted newcomer. I suppose this might be true; however, it seems unlikely to me, and parsimony (the assessment of such likelihoods) is an important part of science. After all, what would have kept the transverse ladybug on sand dunes and pediment slopes to begin with? My own suspicion is that Sloggett and Majerus' model makes more sense of the transverse ladybug. If we think of these arid microenvironments as alternate, temporary habitats originally, then it makes sense that the transverse is able to persist in these habitats that are still in some way unwelcoming to the seven-spot. After all, the transverse is a western species, and the west is often a land of aridity and heat. The seven-spot, by contrast, is at home in more humid, lush environments (think "England"), such as those we find in eastern North America.

A close relative of the transverse, the nine-spot ladybug, may tell a similar story. Like the seven-spot, it is a member of the genus *Coccinella*, most of which are western in North America. As Dobzhansky was first to note, the nine-spot is the only species in the genus that is native in the southeastern United States, and it is in the east that the nine-spot ladybug is declining in the wake of seven-spot and Halloween newcomers. In Alberta, by contrast, the nine-spot appears to be persisting in low numbers, especially on the prairies, and it would be wonderful to know what sorts of environments it is breeding in. Personally, I think it will show the same pattern as the transverse. In the final stages of preparation of this book, I found two nine-spots in scurf pea on the edge of a sand blowout in the stabilized sand dunes north of Purple Springs, Alberta, and a few days later, I found more nine-spots on scurf pea at the edge of an active sand dune near Burstall, Saskatchewan. The connection between the now-uncommon nine-spot and transverse ladybugs, and sand dunes, seems obvious and strong to me.

In the case of the nine-spot, it seems to me that its survival in the west and not the east does indeed support the notion that this species has retreated to ancestral habitats, but in a broadly geographical sense, not in a local sense. It is worth noting, however, that Al Wheeler and Richard Hoebeke argue that the almost complete disappearance of the nine-spot from the northeastern United States might or might not have been caused by the seven-spot. It might also have been due to changes in land use, declines in aphids, parasitism or disease. I think this is entirely plausible, at least in the sense that the seven-spot may have been able to completely replace the nine-spot in a landscape without much undeveloped land.

Ladybug coexistence is clearly possible ecologically, and transverse ladybugs coexist with Halloween ladybugs and seven-spot ladybugs over vast areas of Eurasia, in a variety of habitats. Victor Kuznetsov, the leading expert on Russian ladybugs, had about ten times as many Halloweens as transverse in his collections when he wrote his book. Kuznetsov also says that the Halloween ladybug shares the broad-leaved deciduous forests with polkadot ladybugs in Russia. As you read the species accounts that follow, notice how many of our species also occur in Europe and Asia, alongside the dreaded "invaders."

So what are we to make of the situation here in Alberta, with the seven-spot ladybug fully established, and the Halloween right on our doorstep, so to speak? Well, let's start with the observation that in nature it is normal for some species to be rare, and not all rare species are destined for immanent extinction. It is also normal for things to change and there is nothing ecologically unusual about rare species becoming common and *vice-versa*. However, remember the prediction that appears at the beginning of this chapter—the immanent

Two habitats in which native Coccinella ladybugs seem to be able to persist in the absence of seven-spots. Top: Above the treeline in the Rockies. Bottom: A sandy blowout on the prairies north of Medicine Hat.

extinction of some 473 native species. Even if you revise that figure (which presumably refers to the entire North American ladybug fauna) to include only the 51 aphid-feeding species (in the tribe Coccinellini) of ladybugs in Canada, the prediction is clearly out to lunch, because not a single one of our ladybugs has disappeared—at least so far as we know. In straightforward common sense terms, we have added two species to our fauna, lost none that we know of, and we have increased, not decreased, our biodiversity (even when we measure diversity carefully, using ecological diversity indices). Our ladybugs, in other words, can't be used as examples of the evils of "alien invaders" unless you simplistically equate ecological change with ecological damage (and yes, many people do).

I think that ladybugs are a wonderful example of how nature can change profoundly without "collapsing." They show us, as do most other examples from invasion biology, that the dangers of "alien invaders" are primarily tied to our emotions, and our economy. We like to think that things were "the way nature intended" when Europeans first arrived here some 300 years ago, and that any changes that befell this version of paradise were, by definition, tragic and regrettable. I have to say that I no longer feel much sympathy with these ideas. I suspect that most people with a background in historical biogeography (and especially a familiarity with the fossil record from the past few thousands of years) feel the same way that I do, since we are deeply aware of how so-called ecosystems or communities disassemble, shuffle their components, and reassemble over time, looking not at all like well-integrated "systems" and much more like independent species moving around and adapting to a constantly changing world, not a constant set of ecological companions.

Recently, a very fine collection of scholarly papers has been published in a mainstream book entitled "Species Invasions: Insights Into Ecology, Evolution, and Biogeography," edited by Dov Sax, John Stachowicz, and Steven Gaines. In this book, the contributors "take a different course" and present what the editors refer to as "results that we suspect will be somewhat controversial": competition has been overestimated in the past, extinction is "idiosyncratic" and impossible to predict, evolution can be very rapid, the damaging effects of genetic bottlenecks have been overestimated, invasions increase diversity rather than decreasing it, and fears of biological homogenization of the earth are without clear support. How about that—the majority of these conclusions are exactly what Theodoropoulos predicted. It seems to me that invasion biology is either on the brink of what philosopher Thomas Kuhn called a "paradigm shift" (wherein the more progressive members of this community are busily reinterpreting their subject matter, rewriting their own history, and trying at the same time to diplomatically avoid the ire of their colleagues the anti-alien crusaders) or an even deeper schism between the anti-alien crusaders and the more neutral observers of nature.

Another recent book, "Conceptual Ecology and Invasion Biology: Reciprocal Approaches to Nature" takes the position that conceptual ecologists can help understand invasions, and in turn, invasions can help to fine tune the principles of conceptual ecology. The book clearly acknowledges the long-standing schism between mainstream ecology and invasion biology, the need to reconnect the two, and the dangers of emotional bias when thinking about newly arrived species. It also returns frequently to the idea that "invasions" and their effects are difficult if not impossible to predict, or to prevent.

Given that the unique characteristics of individual species are the key to understanding most invasions, let me finish this discussion with three more comments about the Halloween ladybug. First, although introduced ladybugs are a bad idea, and they are lousy biocontrol agents for aphids, some American entomologists still see value in the Halloween ladybug, whereas the seven-spot makes no difference whatsoever to aphid numbers. Thus, the most hated ladybug in North America may well be the species most deserving of the title, "Gardener's Friend."

Second, the Halloween ladybug is hard to live with, at least for people. They congregate in houses, they smell, they bite people, and they can even affect the taste of wine, as they have in such places as Ohio, New York, Pennsylvania, Indiana, and Ontario. The beetles tuck into the grapes in the fall, both for shelter and for food, and it is difficult to get them out before the grapes are made into wine. Apparently, the resulting wine does not taste like alkaloids, but instead a bit like peanut butter, oddly enough. So again, it is clear that the importation of this beetle was a bad idea, and that applied entomologists should avoid this sort of mistake in future.

Third, it seems clear that the Halloween has a more profound effect on the composition of the ladybug fauna than the other newcomers. Even if we ignore the fact that anti-alien fears were misplaced with respect to the seven-spot, perhaps the Halloween is a scarier beast. It forages not only in low vegetation, but also in trees, and it seems to be responsible for at least local extinctions ("extirpations") in the eastern United States. I still think it is important to remember that we do not have published studies in which good naturalists have searched for native species in all possible habitats after the establishment of the Halloween, but I'll also admit that this is still not sufficient justification for believing that the native species are out there somewhere, hiding. As I was taught, "absence of evidence is not evidence of absence." Of course, it is difficult to picture just what "evidence of absence" should look like, but the fact is we haven't searched as thoroughly as we could have. Here in Alberta, we are in a good position to watch our ladybugs carefully if and when the Halloween "invades" But will it? The answer lies not in some general theory of biological invasions but in our understanding (or lack thereof) of the biology of this particular species.

The closest established populations of Halloween ladybugs are on the west coast. They seem to like the warm, humid climate there, but will they do well in our cold, dry winters? David James, a professor at Washington State University, thinks that eastern Washington is too arid for the Halloween (in an information pamphlet published by his university), and if that is true, then Alberta is probably even more so. In Texas, I have noticed that the Halloween is only moderately common in the dry southern tip near Mission and McAllen but much more abundant in the more humid parts of the state, around Houston for example. The species does, however, do well in Siberia, and for that reason, most ladybug experts predict it will do well here also. It may be, however, that the original stock that was used for introducing this species to North America was adapted to more southerly, humid areas, in which case the species might or might not adapt to northern conditions here. Only time will tell.

Another factor might also come into play. At the 2005 meeting of the Entomological Society of America, two of the research presentations were devoted to the study of a sexually and socially transmitted fungus (*Hesperomyces virescens*) that infects the Halloween ladybug and was discovered in the late 1990s. Christine Nalepa and her colleagues are the primary researchers in this area. Will this fungus spell the end of the Halloween's success in North America? Will it affect only the Halloween? Again, only time will tell.

We are, therefore, in an interesting position, at an interesting point in time. Coincidentally, the British are just a few years ahead of us. With very good identification resources, and an established network of ladybug reporters and a good national database, they are watching with interest as the Halloween spreads across the British Isles, after arriving in 2004. Mike Majerus is the coordinator of the Harlequin Ladybird Survey, and you can follow the progress of this project at www.harlequin-survey.org. Mike, along with Vicky Strawson and Helen Roy, have also published a paper outlining the reasons they are apprehensive about the arrival of the Halloween ladybug.

If you have read this far, you are probably in a position to help here as well, at least with respect to clarifying our understanding of what is going on. I hope that enough naturalists get to know our Alberta ladybugs and begin to keep records of what they see. I keep such records myself, and I'm glad to help anyone else add their data to the communal pool. With the co-operation of such institutions as the E.H. Strickland Entomology Museum, at the University of Alberta, and the Alberta Natural Heritage Information Centre, we can follow the British lead and document our ladybugs as the fauna changes. Of course, introduced species are not the only factors that might affect the ladybugs of Alberta. Other things, such as climate change, changing land use patterns, and interactions with changes in the plants and other animals may also play a role. Without data, we will never know, and without educated naturalists, we will never have the data. With that as a pep-talk, let's proceed to the most enjoyable subject of all, the species-by-species treatment of the ladybugs of Alberta.

The lugubrious ladybug—a rarely seen gem from the sandiest prairies in southeastern Alberta.

5
The Lesser Ladybugs of Alberta

In this book, I have divided the ladybug family into two broad groups. The first is the "lesser ladybugs," so named because most are generally small, rare, and ecologically unlike their larger cousins. So far, they have been of interest primarily to beetle collectors and taxonomists, although some species have been used for biocontrol in the United States.

Technically, these lesser ladies belong to the subfamilies Sticholotidinae and Scymninae. Some are just as colourful and cute as the "larger ladybugs," but none are familiar to the average person. Only one of our lesser ladybugs exceeds four millimetres in length, and most are between two and three millimetres long. Thus, the size range of the lesser ladybugs overlaps that of the larger ladybugs. This arrangement was inspired by the traditional distinction between "micro moths" and "macro moths," a division that is also based on taxonomy, not size (although most "micros" are small and most "macros" are large). If lepidopterists can live with such an arrangement, perhaps ladybugsters can too.

I have to admit that, before working on this book, I had very little interest in the lesser ladybugs myself. In fact, I strongly considered leaving them out altogether, because there are so many of them and they are so rarely encountered. In the end, however, I decided that it would be better to treat them alongside their larger cousins for the sake of completion, and now I'm glad I did. They are mysterious, poorly known beetles and perhaps increasing their profile might result in interesting discoveries and new information coming to light. I am also inspired by the British approach to ladybug appreciation in which all ladybugs, no matter how small, rare, or colourless, are treated equally. Searching for lesser ladybugs is a great example of field entomology at its most refined—they don't simply pop out at you the way larger ladybugs do (or butterflies, damselflies, or tiger beetles for that matter). You have to search carefully and skilfully, and it can be remarkably rewarding when you find a species you haven't seen before.

Subfamily Sticholotidinae

TRIBE MICROWEISEINI

These small ladybugs feed on scale insects and at least some over-winter as full-grown larvae rather than as adults. Since we have only one species here, we really don't get much of a feel for this diverse group of rather un-ladybug-like ladybugs.

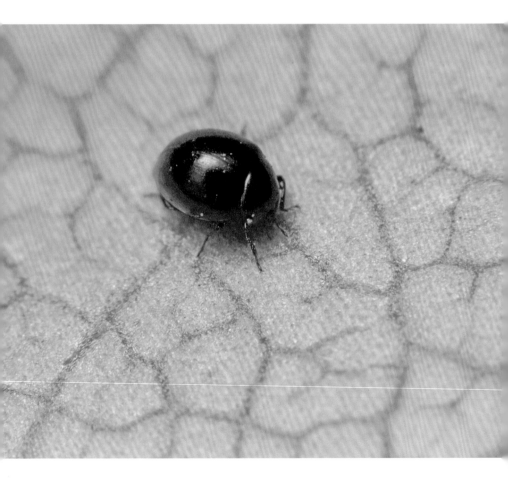

Micro Ladybug

MICROWEISIA MISELLA (LeConte) *("MIKE-row-VYSE-ee-ah miss-ELL-ah")*

"Look at this ladybug!" says author John,
"You gotta be kidding," most readers respond.

THE NAME
"Micro ladybug" seems obvious for this beetle (although one friend of mine calls it "micro-wee-wee-weisia"). Micro means small in Greek, and J. Weise was a beetle scientist in the early 20th century. *Misella* comes from Latin, meaning small, poor, and wretched.

IDENTIFICATION
1.0–1.5 mm, entirely black (or brown when newly emerged), shiny, and only very slightly hairy. Not many other beetles, of any sort, are this small, and if they are, they are generally some other shape or colour. Note that American hairy ladybugs are noticeably more hairy, with lighter-coloured legs, and a less pointy face and butt. If you get a close-up view, another way to tell a micro ladybug is by the conical and pointed (not wedge-shaped and expanded at the tip) last segment of the maxillary palp (the mouth part "feeler" that is shorter than the antennae).

NOTES
Getting to know such a minute beast is, to me, a treat. Once you can recognize the micro ladybug (and even a 1.5 mm all-black ladybug still elicits "aww, how cute" from some people), you begin to feel in your heart that it is truly possible to identify even the smallest of creatures. Gerry Hilchie told me that he had found more than 25 micro ladybugs on the bark of alder trees in the river valley in Edmonton, and I have now found them there as well. I have also found the species on chokecherry, balsam poplar, and mountain ash bark. I weighed one too, by the way, at a whopping 0.0007 grams (i.e., a little over half a milligram)! The micro ladybug is widespread across southern Canada and most of the United States excluding the southwestern deserts.

Subfamily Scymninae

The "scymnines" comprise a huge, diverse group of small ladybugs that generally feed on mites and scale insects, although the larger members of the group do take aphids. Why they are so rare compared to the larger aphid-eating ladybugs is anyone's guess—some are so rare, I have not seen a single specimen. They don't generally appear at wash-up, and they are rarely seen in gardens. Often, I have wished that the early collectors had indicated how they got their specimens when writing their data labels. On the other hand, by carefully scanning the bark of smooth-barked trees such as alder, cherry, and young poplars, I have found some species. An even more productive technique is to use a sweeping or beating net in small shrubs or herbs, especially sage and other small shrubs. Grasses and trees don't produce many of these beetles, but I suppose you never know where some might be hiding. Notice as well that many of our species are confined to the prairies in Alberta, while others are found only in the mountains. Most Albertans with an interest in ladybugs have lived in Edmonton, and the area around Edmonton is not very good for lesser ladybugs—perhaps this has contributed to the notion that they are rare and obscure beetles.

Most of our species can be identified by their colour and size, despite the long history of taxonomic confusion that has surrounded the group. For those who would like a deeper understanding, Robert Gordon's 1976 review, and his 1985 monograph, are the places to go for more details.

A typical, diminutive member of the eastern, yellow-legged race of the American hairy ladybug.

TRIBE STETHORINI

American Hairy Ladybug

STETHORUS PUNCTUM (LeConte) *("STETH-orr-uss PUNK-tum")*

When searching for tiny American hairies,
Try sweeping a net on the wide open prairies.

THE NAME
Stethorus appears to come from the Greek, meaning "breast margin," probably referring to the prosternum, a shelf-like plate that extends under the mouthparts of these beetles. *Punctum* is also Greek, meaning "with small punctures." Since these punctures bear obvious hairs, I have coined the name American hairy ladybug.

IDENTIFICATION
About 1.5 mm. Black, covered with slightly curved pale hairs, and with a deep postcoxal arc. Can be confused with the micro ladybug (which is less hairy and has darker legs), the lacustrine ladybug (which is noticeably larger), and the angular ladybug (which is more elongate).

NOTES
There is one record of the eastern subspecies *Stethorus punctum punctum* from Alberta, a yellow-legged specimen collected in Medicine Hat in 1964 by John and Bert Carr. All other Alberta specimens are typical of the western race *Stethorus punctum picipes*, with brown legs and a deeper postcoxal arc. These are known from Fort McLeod, Grand Forks, Medicine Hat, North Pinhorn Grazing Reserve, and Taber. Since two subspecies cannot exist in the same place, I assume that the western subspecies is the resident race in southern Alberta, and that the eastern specimen was either a displaced individual or a colour variant with western parentage. Members of this genus feed on spider mites (family Tetranychidae), and have been called "spider mite destroyers," but that name makes me laugh so I don't use it here.

Newcomer Hairy Ladybug

STETHORUS PUNCTILLIUM Weise *("STETH-orr-USS punk-TILL-ee-um")*

Each time you find Stethorus prowling on bark,
It's your duty to look at its postcoxal arc!

THE NAME
Punctillium is Greek and refers to the tiny punctures on the pronotum and wing covers of this beetle. I have distinguished the newcomer hairy ladybug from the American hairy ladybug with no intention of denigrating the newcomer because of its origins.

IDENTIFICATION
About 1.5 mm. Black, covered with slightly curved pale hairs, and with dark legs and a shallow postcoxal arc.

NOTES
This is an introduced species that came accidentally to North America from Europe in the 1950s. Perhaps it is established here, perhaps not. In Taber in 2001, Ted Pike found a single female beetle that might be this species. The postcoxal arc of the specimen is relatively shallow, but the punctures on the head and pronotum appear to be equal in size, not "subequal" as expected. It is a female, so dissection of the genitalia was not able to resolve its identity. To put our hairy ladybug records in perspective, this genus was reported from Manitoba for the first time in 2001, by Ian Wise and Bill Turnock. So, we are not far behind in our understanding.

TRIBE SCYMNINI

Mealybug Destroyer

CRYPTOLAEMUS MONTROUZIERI Mulsant
("KRIP-toe-LEE-muss mon-TROO-zee-ERR-eye")

On entering greenhouses through airlock foyers,
Be always on guard for mealybug destroyers.

THE NAME
Crypto means hidden, *laemus* means neck (both are Greek). The species name honours Xavier Montrouzier, a French born entomologist who described many beetles, other insects, and plants in New Caledonia in the 1800s. And yes, the English name makes me laugh. Why not just "mealybug eater"? I suppose "destroyer" probably sells more ladybugs to more greenhouses.

IDENTIFICATION
About 4.5 mm. Like a big, bright *Scymnus* ladybug—dark and hairy, with a rusty orange head, pronotum, and tips of the wing covers.

NOTES
This is a biocontrol ladybug in Alberta that is sold for use in greenhouses, where it feeds on various sorts of scale insects, including "mealybugs." It does not occur in the wild here. This species was originally brought to North America from Australia for this purpose and is now established in the warmer parts of California and Florida. It is highly unlikely that this species could survive outdoors in Alberta, but you might encounter escaped individuals from time to time. I have included it here mostly for anyone who might see one in a greenhouse and wonder what it is.

Twice-stained Ladybug

DIDION PUNCTATUM (Melsheimer) *("DIDD-ee-on punk-TATE-um")*

Twice stained ladybugs, smaller than twice-stabbed,
Are very rare and not at all as likely to be nabbed.

THE NAME
Didion is a very puzzling name, which may be the name of a person, or a made-up word. *Punctatum* is Latin, referring to the small punctures on this beetle. My choice of English name was inspired by the established name for member of the genus *Chilocorus*, the twice-stabbed ladybugs.

IDENTIFICATION
About 1.5 mm. Elongate and brown or black, usually with a single rusty spot on each wing cover, but may be unmarked (at least elsewhere—I have not seen such specimens from Alberta).

NOTES
Widespread in the United States and Canada. May or may not feed on spider mites, and may or may not be found on tree bark, like its angular relative. I look forward to finding this beetle in person, but I also welcome any records that others might have of the species in Alberta.

Angular Ladybug

DIDION LONGULUM Casey *("DIDD-ee-on lonn-GEW-lum")*

Look there on the bark for the angular lady,
I find them in sunlight, rather than shady.

THE NAME
Longulum comes from the Latin, and means elongate but small. "Angular ladybug" draws attention to its distinctive outline.

IDENTIFICATION
About 1.5 mm. Black. Much like the immaculate form of the twice-stained ladybug but with finer punctures on the wing covers, and a more rounded shape (although still noticeably more elongate than the American and newcomer hairy ladybugs). I always notice the angular shape of the pronotum, which does not form a smooth continuation of the curve of the wing covers.

NOTES
A widespread western species in Canada and the United States that is probably found throughout Alberta. May feed on spider mites, at least in part. I have found it twice on the bark of young balsam poplar trees (making it part of an ecological threesome including the twice-stabbed and micro ladybugs as well). In some ways, it is easier to distinguish from other small black ladybugs in the field than it is under the microscope, because its overall body shape is more apparent at a distance. It truly is more elongate.

Apicanus Ladybug

SCYMNUS APICANUS Chapin *("SKIM-nuss ape-ih-KANN-uss")*

The apicanus ladybug
Can only give a tiny hug

THE NAME

Scymnus is a cute word that means a cub or whelp, and is Greek. *Apicanus* refers to either a so-called notorious epicure of ancient Greece, or a Latin book on cooking—either way it probably implies that this ladybug is a glutton. Since it's an easy word to pronounce, I think it makes a good English name as well.

IDENTIFICATION

2.5–3.0 mm. Head and pronotum yellowish brown (except for a rectangular dark patch in the middle at the rear of the pronotum that may extend almost to the front), as well as a narrow strip at the tip of the wing covers. Otherwise black and finely hairy. Apicanus ladybugs are, on average, smaller than paracanus ladybugs, with pronotal punctures noticeably smaller than those on the wing covers, assuming you have the beetle under a good microscope.

NOTES

Widespread east of the Rockies, with our subspecies (*S. a. borealis*) occurring in Alberta, Saskatchewan, Manitoba, and Colorado. I have found apicanus ladybugs on roadside sunflowers (common sunflowers, *Helianthus annuus*, not the commercial sort) a stone's throw east of Alberta north of Burstall, Saskatchewan, but I am not at all sure if this is a typical place to encounter this species. At another environmental extreme, Gerry Hilchie found this species in the Birch Mountains, which lie between Fort McMurray and Wood Buffalo National Park in Alberta's northwest.

Paracanus Ladybug

SCYMNUS PARACANUS Chapin *("SKIM-nuss pare-ah-KANN-uss")*

The only thing that beats a paracanus is two of a ladybug kind,
But what does paracanus mean, or is it merely a rhyme?

THE NAME
Paracanus, according to my classicist friend Selina Stewart, is an oddly constructed name that perhaps should have been "*parapicanus.*" *Para* means beside, and it is likely that the intent was to indicate that this species takes its place "beside" *apicanus* in the general scheme of things.

IDENTIFICATION
About 2.6–3.0 mm. Much like the preceding species but with the elytral punctures not distinctly larger than punctures on the pronotum, and the elytra not as shiny as that of the apicanus ladybug.

NOTES
Widespread with a similar range to the preceding species, including the fact that both have western subspecies that look alike and share a geographic range, suggesting that similar evolutionary selection pressures are acting on both ladybugs to produce what is called convergent evolution. Our subspecies is *S. paracanus linearis*.

Opaque Ladybug

SCYMNUS OPACULUS Horn ("SKIM-nuss oh-PACK-you-luss")

Of course it's opaque, I hear you say
But opaque here means shady, at least for today

THE NAME

Opaculus comes from Latin, and means dark, obscure, or shady, presumably referring to the overall colour of this beetle. "Opaque" is a semi-accurate translation of the name into English, and I like the sound of it.

IDENTIFICATION

2.4–2.7 mm. Black except for the sides and front edge of the pronotum and rear third of wing covers, which are rusty orange. Almost identical to fake-opaque ladybugs, which are smaller and have complete postcoxal arcs.

NOTES

Ranges from Medicine Hat and the British Columbia interior south to Utah and Colorado. This appears to be another dryland, prairie species, and it should be searched for by sweeping various plants in southeastern Alberta. In other words, we know very little about this beetle.

Fake-Opaque Ladybug

SCYMNUS POSTPICTUS Casey *("SKIM-nuss post-PICK-tuss")*

Postpictus, post-dictus, predicting the past
It lives in the prairies, and there it will last

THE NAME

Postpictus means "painted after" in Latin, and probably refers to the light tips of the wing covers. Since it looks so much like the opaque, I coined the rhyming name "fake opaque," which also indirectly pokes fun at many scientific names that begin with *pseudo*, meaning "false," a silly way to characterize living things.

IDENTIFICATION

1.8–2.2 mm. Almost identical to the opaque ladybug, but smaller, and with complete postcoxal arcs (as with all members of the subgenus *Pullus*). The fake-opaque is also more variable.

NOTES

Despite its similarity, this species is in a different subgenus (*Pullus*) from the preceding species (*Scymnus*) and, thus, is another example of convergent evolution in pattern. Fake opaque ladybugs range from the Canadian prairies and the British Columbia interior south to Nevada and Utah. The only ecological information I have for this species in Alberta comes from a single record: I caught one by sweeping prairie vegetation (including small sages) in a coulee northeast of Manyberries, in late July. As with many beetles caught by sweeping, I wished I could go back and find out what it was doing before I knocked it off the plant! Gerry Hilchie found another at the Ghost Dam, washed up on shore, and again away from its breeding habitat.

Carr's Ladybug

SCYMNUS CARRI Gordon *("SKIM-nuss KARR-eye")*

Gordon remembered the work of F. Carr
Who, as far as I know, rarely went to the bar

THE NAME
The name honours Frederick S. Carr, Alberta's preeminent pioneer beetle authority.

IDENTIFICATION
1.7–2.2 mm. Black wing covers, orange head and pronotum. This is the only *Scymnus* in Alberta with an entirely orange pronotum.

NOTES
Known entirely from Alberta so far, this is one of two ladybug species for which the type locality (the place where the type specimen was collected) is Medicine Hat (the other is the Hercules ladybug). Other species of ladybugs have also been named from Medicine Hat, but these names are now considered invalid. John and Bert Carr, along with Ted Pike, have also found this species in southern Alberta. Ted's single record came from Taber, while the Carr's single specimen came from near Carbon, the most northerly record to date.

Uncus Ladybug

SCYMNUS UNCUS Wingo *("SKIM-nuss UNK-uss")*

Uncus means hook, and hooked you will be
When you try to find uncus on the bald-butt prairie

THE NAME
Uncus means a hook in Latin, no doubt referring to part of the genitalia (probably the tip of the basal lobe of the adeagus, if you must know).

IDENTIFICATION
About 2.3–2.4 mm. A lot like the apicanus and paracanus ladybugs but smaller and with complete postcoxal arcs. Slightly larger than the Diamond City ladybug but probably indistinguishable except by dissection.

NOTES
In Alberta, this species is known only from Medicine Hat, possibly as an accidental occurrence. So far, my own searches (and I do spend time in Medicine Hat each summer) have been fruitless, and Robert Gordon's map shows this species much farther to the east and south in North America. It is my opinion that this species and the Diamond City ladybug will only make sense once we have a good number of specimens from southern Alberta, and someone sits down and patiently dissects the males.

Diamond City Ladybug

SCYMNUS AQUILONARIUS Gordon *("SKIM-nuss ah-QUILL-oh-NARE-ee-uss")*

The northerly ladybug from Diamond City
Is rarer than hen's teeth, and that's a dang pity

THE NAME
Aquilonarius means northerly or belonging to the northern wind. As for the English name, I like the sound of "Diamond City" so I suggest we use it!

IDENTIFICATION
About 2.1 mm. Much like the uncus ladybug (with which it shares complete postcoxal arcs), but smaller and probably indistinguishable without dissection of the males.

NOTES
This species is known for sure from only one specimen, collected at Diamond City, north of Lethbridge. I doubt that there is anything ecologically special about the area around Diamond City, so the species should be searched for throughout southern Alberta. This species and the uncus ladybug are so close in external appearance that I doubt we will understand the distribution of either until we have much larger collections of specimens.

Lacustrine Ladybug

SCYMNUS LACUSTRIS LeConte *("SKIM-nuss lah-KUSS-triss")*

This oddball lady looks oh so very
Much like a giant American hairy

THE NAME

Lacustris means lake in Latin, and indeed I have usually found this beetle somewhere near water, although I doubt that this is an essential component of its habitat.

IDENTIFICATION

2.1–2.5 mm. Mostly black with fine hairs all over, with dark orange and black legs and antennae. Males have a dark yellow face and front angles of the pronotum.

NOTES

This is our most commonly encountered *Scymnus*, and the only one that I can really say I am "familiar" with. The species is widespread in Canada and the northern half of the United States, although perhaps not reaching the east coast. One old specimen from Medicine Hat is labeled "on *Sagittaria*," referring to the arrowhead, a semi-aquatic plant, but this is clearly not its normal habitat. When I first started finding lacustrine ladybugs, I'll admit that I mistook them for American hairies, not noticing the difference in size. Lacustrine ladybugs seem to be most common on small shrubs, and I have found them with a sweep net on cherry, willow, wolf willow and the like. Often, the net will also contain black aphids almost exactly the same shape and size as the beetles.

Ornate Ladybug

NEPHUS ORNATUS (LeConte) *("NEEF-uss orr-NATE-uss")*

Nephus ornatus is about as ornate
As a bit of dried bird poop on a wrought iron gate

THE NAME
Nephus is Greek for cloud, possibly referring to the markings on the wing covers, and *ornatus* is Latin meaning decorated, again probably referring to the wing covers. "Ornate ladybug" may be an overstatement, but it's a nice translation of the scientific name.

IDENTIFICATION
1.6–2.0 mm. Each black wing cover with a prominent light orange somewhat peanut-shaped spot.

NOTES
Transcontinental in Canada and the northern United States. Our subspecies is *N. o. naviculatus*, and it differs from the eastern subspecies in which the spots are separate from one another. There are specimens in the E.H. Strickland Entomological Museum from Edmonton, and the Ghost Dam (the latter almost certainly at wash-up). Vanessa Block found it at Colin-Cornwall Lakes Wildland Provincial Park. In New Brunswick, it has been found in bogs and fens and in sphagnum moss.

Farmer's Ladybug

NEPHUS GEORGEI (Weise) ("NEEF-uss GEORGE-eye")

Georgei, porgei, puddin' and po
Who was that George? I really don't know.

THE NAME
Most Alberta beetles with the name *georgei* are named for George Ball, the University of Alberta professor, but since this species was described in 1929, it must honour another George, whose identity I have not been able to determine. Alternatively, it may refer to the Greek word "George," which means farmer. Either way, the English name is justifiable.

IDENTIFICATION
1.5–1.7 mm. A smaller and darker beetle than the ornate, lacking any expansion of the dark line down the midline of the wing covers. The light spot is more reddish.

NOTES
This species is apparently about as widespread as the last, but we have precious few records for Alberta or for anywhere else, for that matter. I suspect that it uses similar habitats to the ornate ladybug and that once we focus in on where these beetles live, we will be able to say much more interesting things about them.

Sordid Ladybug

NEPHUS SORDIDUS (Horn) *("NEEF-uss SORR-did-uss")*

How do we look for a ladybug sordid,
When none of us know where the records are hoarded?

THE NAME
Sordidus means "dirty" in Latin, and probably refers to the overall colouration of this beetle. I think that combining a word like "sordid" with the word "ladybug" is fun, so that's the English name I choose.

IDENTIFICATION
1.5–2.0 mm. Brownish, not black, with a large yellowish brown area on each wing cover, which may be large enough to cover the entire wing cover.

NOTES
Widespread from Alberta south throughout the western United States. I've never seen an Alberta specimen, and I'm not at all sure where to search for them. It is likely that learning to find the sordid ladybug will be closely tied to learning to find the farmer's ladybug, since it is the ecology of this genus that is such a mystery right now. The best clue I personally have right now is a single specimen collected by John and Bert Carr on the prairie near Hays (south of Brooks), on May 21, 1978.

TRIBE SELVADIINI

Tinytan Ladybug

SELVADIUS NUNENMACHERI Gordon
("sell-VAY-dee-uss NOO-nenn-MACK-err-eye")

Tiny tan from tinytown, hiding in the prairie ground,
Small, elongate, leather brown, that's the one that we have found!

THE NAME
Selvadius probably refers to forests—the word *selvadi* means forest in Romansch and is derived from the Latin word *silva*, often used in the construction of animal names. Frederick William Nunenmacher was an amateur entomologist with an interest in ladybugs (and his name comes up again under the mimic ladybug). He worked mainly in California and donated some 15,000 ladybug specimens to the California Academy of Sciences.

IDENTIFICATION
About 1.5–2.5 mm. Very small, and entirely tan in colour. Relatively elongate as well.

NOTES
This tiny tan creature was "discovered" for Alberta when I borrowed two specimens from Gerry Hilchie's collection, both collected by John and Bert Carr. The first was labeled "*Selvadius rectus*" and came from near Acme. *Selvadius rectus* is a species of the southwestern United States, and was unlikely for Alberta. The second was labeled "*Diomus debilis*" and came from near Hays. This second specimen is probably the reason that the name *Diomus debilis* appears for Alberta in the published checklist of Canadian beetles. Both, on close examination, proved to be *S. nunenmacheri*, a species formerly known from Colorado and Wyoming—a bit closer to home. It's great to have it all cleared up, and in the process we also gained another tribe of ladybugs for the province.

TRIBE HYPERASPIDINI

Mimic Ladybug

HYPERASPIDIUS MIMUS Casey *("HIPE-err-ass-PIDD-ee-uss MINN-ih-muss")*

The mimic is tiny, the mimic is striped,
But its mimetic nature is perhaps over-hyped.

THE NAME
Hyper means above or over in Greek, and *aspis* means a small shield—a combination that could apply to the upper parts of the exoskeleton of almost any insect. *Mimus* comes from Latin, meaning mimic, although it is not clear what this tiny striped creature might be mimicking. There are many small striped beetles and plant hoppers in the same environment, and it may well be that some of these actually mimic the mimic ladybug.

IDENTIFICATION
1.4–1.8 mm. This is a typical striped *Hyperaspidius* ladybug. As with all members of the genus, males have a yellow head and more yellow on the pronotum, while females have a dark head and less yellow on the pronotum. This species and the next are both small as adults, but they are tough to tell apart and may require dissection for confirmation of their identity. I get the impression that the markings of the mimic are browner and less contrasting than those of the vittate, but I haven't seen enough Alberta specimens to say for sure.

NOTES
Ranges from the grasslands of the Prairie Provinces south to Utah and Colorado. Adults of this genus probably feed on scale insects, and they are difficult to catch with a sweep net since they live very close to the ground. Our Alberta populations were originally named *H. carri* by F.W. Nunenmacher in 1948 (to honour F.S. Carr), but this name is no longer in use.

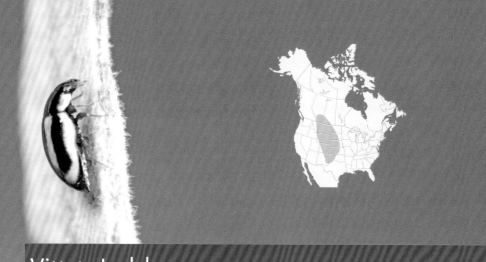

Vittate Ladybug

HYPERASPIDIUS VITTIGERUS (LeConte)
("HIPE-err-ass-PIDD-ee-uss vih-TIDGE-err-uss")

The vittate ladybug down in the Hat,
Is deeply obscure, and that is that.

THE NAME
Vittigerus comes from Latin and means ribbon. It probably refers to the striped colour pattern. "Vittate" means the same thing.

IDENTIFICATION
1.5–2.0 mm. Much like the mimic, and quite likely impossible to identify without dissection, but potentially darker and with stronger contrast in the markings.

NOTES
Ranges from southeastern Alberta east to Manitoba and South Dakota and south to New Mexico. Assuming our records are all correctly identified, this is the most commonly encountered member of the genus in Alberta, especially near Medicine Hat. I have found this beetle numerous times in the sparse vegetation on the edges of grown-over prairie sand dunes, such as the ones near Empress, Hilda, and Pakowki Lake.

Well-marked Ladybug

HYPERASPIDIUS INSIGNIS Casey *("HIPE-err-ass-PIDD-ee-uss inn-SIGG-niss")*

Insignis, well-marked, and tres hard to find,
I'll smile when I see one, deep in my mind.

THE NAME
Insignis does indeed mean "well marked" and it comes from Latin.

IDENTIFICATION
2.3–3.2 mm. This and the Hercules are noticeably larger than the preceding two species. Well-marked ladybugs with extensive dark markings are easy to identify. Those with the more typical striped pattern on the wing covers are recognizable by size, and by their pale yellowish pronotum with faint markings if any.

NOTES
Ranges from southeastern Alberta to Oklahoma. I have never seen a specimen, let alone encountered the species in the wild.

Hercules Ladybug

HYPERASPIDIUS HERCULES Belicek
("HIPE-err-ass-PIDD-ee-uss HERK-you-leez")

This bug, indeed, is like the great Hercules,
After all, it's bigger than most kinds of fleas.

THE NAME
Hercules was, of course, a figure in Greek mythology, and his name implies large size and strength. For a giant among tiny relatives, the name is appropriate.

IDENTIFICATION
2.1–4.0 mm. Our second large *Hyperaspidius*, and big individuals are the largest of the genus. Typically striped, with a strongly marked pronotum. Some small males (about 2.1 mm long) are almost identical to large male vittate ladybugs, but the Hercules males should have an incomplete rather than complete post-coxal arc.

NOTES
Ranges from southern Alberta and Manitoba to California and Colorado. Joe Belicek named the species and designated a male collected by F.S. Carr in 1932 in Medicine Hat as the type specimen. I found one on a Virginia creeper at my in-laws' acreage just upstream near Redcliff and another on a grassy sand dune near Empress (in a pitfall trap). Ted Pike found a single female at Grand Forks, Alberta, where the Oldman River meets the Bow.

Unnamed Ladybug

HYPERASPIDIUS SP. *("HIPE-err-ass-PIDD-ee-uss sspUH")*

It's tiny and dark and a real big shame,
That this thing's in the book but it don't have a name!

THE NAME
Since I'm not sure what it is, the name of this ladybug is still a mystery.

IDENTIFICATION
Our single specimen is 2 mm long, and all black, with a pale mark along the margin of the shoulder and a pale spot near the tip of the wing covers.

NOTES
I wish I knew to call this thing! There is a single specimen in Gerry Hilchie's collection, labeled " Tp. 27 Rge. 17/ w. 4 Mer. Alberta/5 VIII 1982/ Lot 2 BF & JL Carr." It was identified as *Hyperaspidius comparatus*, but that species is never this dark in colour. The specimen is a female, and it resembles a number of species from the United States, some of which are unknown as females. So, we'll just have to wait until someone catches a male before we can identify this creature for certain. And where is " Tp.27 Rge.17 w.4"? Why, it's along the valley of the Red Deer River near Dorothy, of course. Please, if you ever become an insect collector, be sure to put actual place names on your labels, along with whatever coordinates you think might pass the test of time!

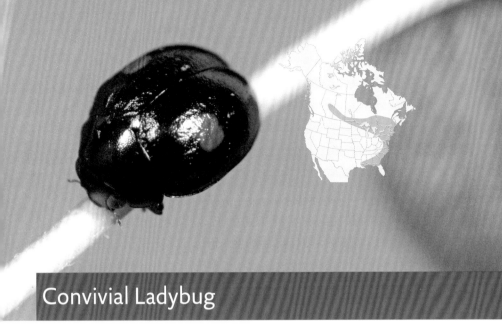

Convivial Ladybug

HYPERASPIS CONVIVA Casey *("HIPE-err-ASP-uss kon-VEE-vah")*

Convivial lady, enjoying good company,
Here I come sweeping and hoping I'll bump any

THE NAME
Conviva means convivial, which in turn means enjoying good company. Perhaps Colonel Casey found a number of them together when the species was first described.

IDENTIFICATION
2.7–3.8 mm. Female all black, with two red spots on the wing covers. Male with a yellow face and front margin of the pronotum.

NOTES
This was a last-minute addition to the list of Alberta ladybugs. On August 28, 2006, I was sweeping some small aspens and saskatoons in a pine forest near Opal when a single female convivial ladybug appeared in my beating net. This species is known from Saskatchewan, so it isn't a great surprise to find it here. Confirming its identity was tricky, however, since there are a few members of this genus with the "twice-stabbed" pattern. The convivial is distinctly flattened on top, as opposed to rounded and dome-shaped. The only ladybugs you might confuse it with in Alberta are the twice-stabbed and once-squashed ladybugs, but the difference in body shape should be obvious if you are careful about it.

Lugubrious Ladybug

HYPERASPIS LUGUBRIS (Randall) *("HIPE-err-ASP-uss loo-GOO-briss")*

Lugubrious lady, you make me so glad,
But your name, in a word, is intended as "sad"

THE NAME
Lugubris comes from Latin, and means dark, gloomy, and sorrowful, much like the English word "lugubrious."

IDENTIFICATION
2.4–3.3 mm. Two orange spots on each wing cover—one central, the other near the tip—which may be joined. Pale border of the wing cover extends from the shoulder to the level of the hindmost spot, to which it may be narrowly connected. Male pronotum is entirely reddish (with a faint half-flower design); female is darker with narrow pale lateral borders.

NOTES
Until I collected a male (with especially extensive light markings) in a pitfall trap on the Empress sand dunes, on June 12, 1984, the only people to encounter this species in Alberta were John and Bert Carr, at various places in southern Alberta, including Lundbreck Falls and the Ghost Dam. Robert Gordon did not list the species from Alberta and was probably unaware of the Carr records, but I am happy to report that it is indeed here. I confirmed this most recently in July 2006, while taking the final photos for the book, when I found a female lugubrious ladybug by sweeping sandy prairie near the Pakowki Lake sand dunes. This is a striking little ladybug, and finding it makes me anything but lugubrious!

Lateral Ladybug

HYPERASPIS LATERALIS Mulsant *("HIPE-err-ASP-uss latt-err-ALL-iss")*

Lateral ladybugs living on sage
Down on the prairies they're all the rage

THE NAME
Lateralis comes from Latin, meaning "the side," and indeed, the markings along the sides of the wing covers are important for recognizing this beetle.

IDENTIFICATION
2.5–3.8 mm. Wing cover margin is orange or yellow-orange along the front half or so. Two orange spots on each wing cover—one central, the other near the tip. The spots can be small or large and may be connected to the pale margin. Usually, the central spot is quite dark and reddish, and the rest of the markings a light creamy yellow.

NOTES
Widespread in the western United States, southern Alberta, and southern British Columbia and accidentally introduced in Florida and Louisiana. It is the most abundant member of the genus. A number of us have found it common on aphid-bearing sage brush on the prairies and, on occasion, in alfalfa as well. Older records of "*H. simulatrix*" from Alberta probably refer to this species. I find it interesting that as you go northward this species becomes less common but more typically red in its markings (rather than white, yellow, orange, or a combination of all three). This may yet prove to be a consistent pattern, worthy of a subspecies name, but I have seen far too few specimens to propose this myself.

Fastidious Ladybug

HYPERASPIS FASTIDIOSA Casey *("HIPE-err-ASP-uss fass-TIDD-ee-OH-sah")*

Fastidious ladybugs, fussy and small,
How can you be when you're a millimetre tall?

THE NAME
Fastidiosa and fastidious both come from Latin, meaning "loathing" but implying that these beetles are exacting and fussy in their requirements.

IDENTIFICATION
2.1–2.7 mm. A bit like the lateral ladybug, but smaller, more flattened, and with yellow spots and markings, including a noticeably elongate middle spot on each wing cover. These beetles are also strongly inclined to play dead when you disturb them. The rear wing cover spots are generally joined to the pale margin, and the front and rear spots may also be joined.

NOTES
Originally, this species was known in Alberta only from Medicine Hat (two old specimens in the E.H. Strickland Entomology Museum). I found it again in the sand hills near Hilda, by sweeping a net through sagebrush. On the same sage plants there were lateral and seven-spot ladybugs. This beetle is also known from much of the western United States.

Curved Ladybug

HYPERASPIS INFLEXA Casey ("HIPE-err-ASP-uss in-FLEX-ah")

Inflexa means curved, and curved means inflexed,
With this bonehead name our wee lady is vexed.

THE NAME
Inflexa comes from Latin and means to bend or curve. It could refer to any number of curved structures on this beetle.

IDENTIFICATION
2.0–2.8 mm. Light stripe extends from shoulder along wing cover margin to join the light spots near the tip. No isolated light spots. Head and front margin of pronotum light in males, dark in females, as is typical for the genus.

NOTES
Found from Opal to southern Alberta, interior British Columbia, and much of the continental United States. This and the lateral ladybug are probably our most common members of the genus *Hyperaspis*—at least there are more records of these two species than for any of the others (although this could easily change). My only personal encounter with this beetle occurred while sweeping pasture sage, scurf pea, and other low plants on a sandy trail in the Hilda Sand Hills. It is a delightfully patterned beetle, and I certainly look forward to seeing more of them. John and Bert Carr found two of these beetles washed up on the shore of the Ghost Reservoir, one of which was much darker than the usual. Two even darker specimens came from Opal and Edmonton, and it may be that the markings are reduced in the northern parts of this species' range, a situation that may eventually require a subspecies name for these beetles.

Postica Ladybug

HYPERASPIS POSTICA LeConte *("HIPE-err-ASP-uss POSS-tick-ah")*

Prairie postica, deep in the sage
I think they have hidden there all through this age

THE NAME
Postica comes from Latin and means "that which is behind." This probably refers to the wing cover spots. Since the name postica is short and easy to say, I think it makes a good English name as well.

IDENTIFICATION
2.3–3.1 mm. Almost all black with a light spot near each wing cover tip. The outside corners of the pronotum are light as well.

NOTES
A west coast species also known from in or near Medicine Hat. This has been interpreted as an accidental introduction (although I think the range might be continuous), but to me this is a less interesting question than whether or not the species can still be found in Alberta. The two known Alberta specimens were collected in 1927 by F.S. Carr and then in 1963 by his son John Carr, suggesting that it breeds in the area. As well, the first specimen was caught "on sage brush," a classic habitat for members of this genus.

Oregon Ladybug

HYPERASPIS OREGONA Dobzhansky *("HIPE-err-ASP-uss orr-egg-OH-nah")*

Oregon ladybugs, in the national park
Protected by wardens, keeping us in the dark

THE NAME
Oregonensis means "of Oregon," and although this is strictly not true, it makes for a nice English name as well.

IDENTIFICATION
2.0–2.5 mm. The most boldly marked individuals are black with a light leading edge to the pronotum, a thin light line in the rear half each wing cover, a small light spot near the tip, and a light margin on the front two-thirds. The most sparsely marked individuals look like the postica ladybug but with the wing cover spots closer to the head.

NOTES
Primarily a Pacific Northwest species, much like the postica ladybug, but known in Alberta primarily from Banff. I have not encountered any of the mountain-dwelling lesser ladybugs, but I suspect that they are found in alpine meadows on short vegetation reminiscent of their relatives' habitats on the prairies. Since insect collecting is not allowed in national parks without a permit, we have very few records of this beetle. It is a shame that the national parks are our least well-known places when it comes to insect life—one might think the opposite should be true. In many places, however, this situation is changing for the better, as park managers realize the value of collaborating with naturalists rather than treating them as a threat. Our only non-Banff record came from wash-up at the Ghost Dam and might well have been a Banff ladybug blown off course!

Blotch-backed Ladybug

HYPERASPIS DISCONOTATA Mulsant
("HIPE-err-ASP-uss DISK-oh-note-ATE-ah")

Disconotata, *a most notable disco*
Is even more rare down in ol' San Francisco

THE NAME
Disco is Greek, meaning disk and probably referring to the wing covers. *Notata* is Latin, and means marked, which is appropriate. "Blotch-backed ladybug" is my way of capturing this notion in English.

IDENTIFICATION
2.3–2.8 mm. The light margin of each wing cover is broken into a shoulder blotch, a middle blotch, and a blotch near the tip. There is also a spot near the middle of the wing cover at the front and one farther back. This is therefore the member of the genus with the most complex markings.

NOTES
An eastern species originally known in Alberta only from Edmonton. As with the postica ladybug, this was an F.S. Carr discovery, way back on June 5, 1920. This beetle was originally considered an accidental introduction, but it has since been found near Spruce Grove by John and Bert Carr in 1985 and in the Caribou Mountains by Gerry Hilchie in 2003. It is clearly one of our more northern lesser ladybugs, and a very pretty beetle as well.

Undulate Ladybug

HYPERASPIS UNDULATA (Say) *("HIPE-err-ASP-uss und-you-LATE-ah")*

The lady waved, and I waved back,
The road was paved, and the cards were stacked

THE NAME
Undulata is Latin and means "waved," a good description of the wing cover markings. Undulate, of course, means the same thing.

IDENTIFICATION
2.3–2.7 mm. Black with a pale yellowish margin to the pronotum, the markings on the edge of the wing covers deeply curved (or broken into three separate blotches), and a light spot in the center of each wing cover. Male has more yellow on the head and front margin of the pronotum. Compared to the fastidious ladybug, the undulate has a more circular central yellow spot, and a more deeply indented yellow edge to the wing covers.

NOTES
Transcontinental through southern Canada and the northern United States. Known from Calgary, Coaldale, Edmonton, Olds, the Wagner Natural Area, and Medicine Hat. I have yet to find it in Alberta, but I did catch one on roadside sunflowers just east of Alberta near Burstall, Saskatchewan. I originally took it to be a fastidious ladybugs, but with only a bit of practice you can distinguish the two quite easily.

Poorly-known Ladybug

HYPERASPIS CONSIMILIS LeConte *("HIPE-err-ASP-uss con-SIMM-ill-iss")*

Yes, it is true, this one is poorly known
Ladybug, ladybug, where have you gone?

THE NAME

Consimilis comes from Latin and means "exactly like." There are at least two other species that LeConte may have been thinking of when he coined the name, but luckily here in Alberta it is easy to identify. I have coined the name "poorly-known ladybug" in the hopes that we will soon learn more about it, and need to change the name.

IDENTIFICATION

2.3–2.7 mm. Wavy light margin on the wing covers, a spot near the front of each wing cover and a long streak toward the tip.

NOTES

A very poorly known ladybug indeed, found in Alberta only at Whitford Lake, originally, from which it was mistakenly described by Theodozius Dobzhansky as a new subspecies of the blotch-backed ladybug, "*H. disconotata canadensis.*" Otherwise, the species is known in Alberta from Edmonton, Valleyview, the Birch Mountains, and Colin-Cornwall Provincial Park. This is clearly a northern ladybug, at least in the western end of its range here in Alberta.

Four-streaked Ladybug

HYPERASPIS QUADRIVITTATA LeConte
("HIPE-err-ASP-uss QUAD-rih-vitt-TATE-a")

Four ribbons glint from its elytral shields,
But alas far too tiny, all I see is fields

THE NAME
Quadrivittata literally means "four ribbons" in Latin but here refers to the markings on the wing covers. "Four-streaked ladybug" is therefore a good English equivalent.

IDENTIFICATION
2.0–2.7 mm. Black with narrow pale wing cover margins and an elongate light streak on each wing cover.

NOTES
May look a bit like the paler Oregon ladybug but with longer pale streaks on the wing covers. Widely distributed (but not at all easy to find!) in the southern Canadian prairies and the western United States. There is also a specimen in Gerry Hilchie's collection from the Birch Mountains (north of Fort McMurray), which suggests that the northern portion of the species' range is especially poorly understood.

Jasper Ladybug

HYPERASPIS JASPERENSIS Belicek *("HIPE-err-ASP-uss JASS-purr-ENSS-iss")*

Black bald beetle, from the big Bald Hills,
Let's go back and look for it, and all relive all the thrills!

THE NAME
Jasperensis means "of Jasper," referring to Jasper National Park.

IDENTIFICATION
1.5–2.0 mm. Entirely brown-black and without obvious hairs, making it the third such species in our fauna, along with the micro ladybug and the tiny black ladybug.

NOTES
Named by Joe Belicek for its type locality (the Bald Hills in Jasper), this is surely one of the few species of organism named from, or for, a national park, since historically such places have been off-limits to entomologists. Peter Kuchar collected the first specimens. It is an alpine species in Jasper and has been found elsewhere: twice in Colorado and once in Wyoming.

TRIBE BRACHIACANTHINI

Pale Anthill Ladybug

BRACHIACANTHA ALBIFRONS (Say)
("BRACK-ee-ah-KANTH-ah ALB-ih-fronz")

Brachicantha albifrons,
Is rarely encountered in gardens and lawns.

THE NAME
Brachiacantha comes from the Greek, meaning short thorn or prickle, in reference to a distinctive spine on the tibia of the front leg. *Albifrons* is Latin, and means white forehead. Because it lives with ants and is lighter in colour than our other anthill ladybug, I have called it the "pale anthill ladybug" here.

IDENTIFICATION
3.5–4.4 mm. Yellow wing covers with two dark spots each, near the outer margin, and a midline stripe that can be either straight and thin, or thick and bulging outward toward the spots, sometimes joining them. The two main spots can also be joined front to back. Males have a light finger-shaped mark extending from the light front edge of the pronotum onto the pronotum for almost half its length. On females the pronotum is mostly black.

NOTES
Larvae in this genus live in the nests of *Lasius* ants feeding on scale insects domesticated by the ants. *Lasius* nests typically have a raised circular mound around the entrance, less than 10 centimetres in diameter. What the adult beetles do is still a mystery. Ted Pike found a female at the Jenner Bridge that is both larger (4.8 mm) and less heavily marked than any others I have seen.

The Lesser Ladybugs of Alberta

Ursine Anthill Ladybug

BRACHIACANTHA URSINA (Fabricius) *("BRACK-ee-ah-KANTH-ah urr-SINE-ah")*

Ursine beetle, living with ants,
Most of the time, we find it on plants.

THE NAME
Ursina is Latin and means resembling a bear, as does ursine.

IDENTIFICATION
2.2–3.0 mm. Black with five light spots on each wing cover—the shoulder spot may be joined to the spots closest to it. Pronotum black with a yellow front margin, more extensively in the males.

NOTES
To be honest, this species puzzles me, and I look forward to learning more about it in the future. Our Alberta ursine anthill ladybugs are apparently smaller than those from farther east (where Robert Gordon gives the size range as 3–4 mm), and this species is also closely related, in a confusing way, to the western *B. uteela*. What we need is a careful comparison of these ladybugs throughout southern Alberta and Saskatchewan, as well as Montana and North Dakota. Until then, however, I suggest we call our Alberta beetles *B. ursina* and hope we are correct.

Ted Pike found many of these ladybugs on snowberry bushes just outside Taber Provincial Park on a fairly steep slope. I searched this habitat on August 23, 2006, and failed to find the beetles. I did, however, find one on the Empress sand dunes in June 1984.

The polkadot ladybug comes in primarily pink, black, and yellow forms—this is the more common pink form.

The Halloween ladybug, a species that may or may not colonize Alberta in the near future.

6
The Larger Ladybugs of Alberta

Not all of the "larger" ladybugs are large, but in general this is the group we think of when we think of ladybugs. Most are colourful, abundant, and widespread, and most feed on aphids in plain sight on low-growing plants. There are, of course, exceptions to all of these generalizations, however. Technically, the larger ladybugs belong to the subfamilies Chilocorinae, Coccidulinae, and Coccinellinae. I suppose the coccidulines (we have one species, the snow ladybug) could have been placed with the lesser ladybugs, since they are indeed small, but they are such oddballs that they don't really look like any of the lesser ladybugs, so I'm inclined to place them here.

The English names I use, by the way, are not exactly "official." Where a name is in use already, I generally use it. However, when no name exists or the existing name is not to my liking, I have felt free to coin new names. Hopefully, they will stand the test of time. I have certainly found that most naturalists want both English names and up-to-date scientific names for the living things they study.

Subfamily Chilocorinae

Most members of this group (the subfamily Chilocorinae and the Tribe Chilocorini) are scale-insect feeders, and although they are medium-sized to large, and usually quite colourful, they have a different overall look compared to the more typical larger ladybugs in the subfamily Coccinellinae. This comes in part from differences in the details of the head and, in most species, from a pronotum that extends well out front on both sides of the head as well. Getting to know the four species in this subfamily here in Alberta is relatively easy, since all but the round black ladybug are reasonably common (and no one can tell the twice-stabbed ladybug from the once-squashed ladybug).

The winter ladybug, in some rather autumnal-looking dry grasses.

TRIBE CHILOCORINI

Winter Ladybug

BRUMOIDES SEPTENTRIONIS (Weise)
("broo-MOY-deez sepp-TENT-ree-OWN-iss")

Ah, winter ladybug, so like a Brumus,
Where do you live? The question consumes us.

THE NAME
There is another genus in the Old World called *Brumus*, and *–oides* means "like," thus "like *Brumus.*" In turn, *brumus* refers to winter, as does *septentrionis*. The name "winter ladybug" is therefore obvious, although there is nothing particularly wintery about this species. It was named from Hudson's Bay, however, a place that feels wintery all year long to most people.

IDENTIFICATION
About 3 mm. Red with a black pronotum, a black stripe along the line where the wing covers meet, and two black spots on each wing cover, which may be absent, expanded, or joined together. This is always an easy ladybug to recognize.

NOTES
These beetles feed on scale insects, and Joe Belicek thought they might be associated with coniferous trees. I have now found them at wash-up in Waterton, as well as on sage and other low vegetation in various places on the prairies. I think of this as a dryland species, with no need for conifers. It does become easier to find late in the season, making the name "winter ladybug" quite sensible. This is a transcontinental species, with three subspecies: ours is the widespread *B. s. septentrionis*, the northernmost of the three.

Round Black Ladybug

EXOCHOMUS AETHIOPS (Bland) *("ex-oh-KOME-uss EE-thee-OPPS")*

Black and round, round and black,
If it's not black and round then you're on the wrong track.

THE NAME
Exo means outside or without, and *chomus* means a mound of earth or heap of rubbish, possibly referring to the first discovered habitat of this beetle (and not likely to mean "rounded shoulder" as some authors have suggested). *Aethiops* means black and scorched, referring to the beetle's colour. I have coined "round black ladybug" for this species.

IDENTIFICATION
3.0–4.2 mm. Another entirely black ladybug, and the largest and most perfectly rounded of those found in Alberta.

NOTES
Found from southern Alberta and Saskatchewan south over much of the western United States, this species does not reach the west coast or Mexico. Like most other members of its subfamily, the round black ladybug feeds on a variety of aphids and scale insects. Some members of this worldwide genus have been introduced to North America for biocontrol purposes, but this is not one of them.

Twice-stabbed Ladybug

CHILOCORUS STIGMA (Say) *("KYE-loh-KORR-uss STIGG-mah")*

*Chilocorus stigma, looks twice impaled,
Sits on the bark, and devours the scales.*

THE NAME
The name "twice-stabbed ladybug" is in wide use. *Chilocorus* probably means full-lipped, refering to the clypeus of this beetle. *Stigma* means a blood spot, although it can also mean "marked" in the broader sense.

IDENTIFICATION
3.8–5.0 mm. Round and black with one red spot on each wing cover, hence the name.

NOTES
The twice-stabbed ladybug is widespread over much of southern Canada from Alberta west, as well as over most of the continental United States (except the far west) and Mexico. These ladybugs are found most often in deciduous trees, especially medium-sized trees (with a diameter of about 10 centimetres) where they presumably feed on scale insects. They also appear at wash-ups, but infrequently, and I get the general impression that they are never terribly common anywhere. The prominent branched spines of the larvae should make them easy to recognize. This is a distinctive, handsome ladybug, and its colouration is almost certainly a sincere warning of its distastefulness. I once tasted the reflex blood of a related species, *C. cacti*, in Florida, and was nauseated for nearly eleven hours—the most severe reaction I have had to any ladybug ever. That beetle was found on an oleander bush, and oleander leaves are dangerously toxic to humans—I really shouldn't have tasted it.

Once-squashed Ladybug

CHILOCORUS HEXACYCLUS Smith *("KYE-loh-KORR-uss hex-ah-SYKE-luss")*

"Three cryptic species," said the good Dr. Smith,
Some have ignored him, thinking "must be a myth!"

THE NAME
Hexacyclus refers to the six ring chromosomes that typify the species. Once-squashed refers to the technique that is used to examine the chromosomes of beetles, which is called a "chromosome squash."

IDENTIFICATION
Exactly identical to the twice-stabbed ladybug, except for the form of the chromosomes.

NOTES
This species was identified by Stanley G. Smith in 1959, on the basis of its chromosomes. It is otherwise identical to the twice-stabbed. This is not simply a matter of one species with variable numbers of chromosomes, as is seen in many other insects. Smith considered the chromosome configuration of this new species to be too profound for them to be oddball twice-stabbed ladybugs. I'll admit that I was reluctant to accept the existence of this cryptic species, thinking that 1959 science might have been mistaken, until I got in touch with my friend David Maddison, who has worked on beetle chromosomes himself. In David's opinion, Smith "knew beetle chromosomes—better than anyone ever has or ever will. ...Until more detailed studies are done (e.g., molecular), I'd go with Smith's conclusions. My gut feeling is that Smith might be wrong—but perhaps in the other direction! That is, there may be more species than he had thought. Just a wild guess." Let's hope that someone does pick up on this, and apply modern DNA-based techniques to this 47-year-old mystery!

Subfamily Coccidulinae

In Alberta, with only one member representing the group, the subfamily Coccidulinae is easy to recognize. Elsewhere in North America the group is more diverse, with four tribes, and quite difficult to characterize (our single species belongs in the Tribe Coccidulini). Because the snow ladybug is relatively small, somewhat dull-coloured, and quite elongate in body shape, those with a beginning interest in beetles may easily mistake it for a leaf beetle. Leaf beetles, by the way, are tremendously diverse in Alberta, and often colourful, and they are fully worthy of attention in their own right (although sadly not well covered by any of the existing field guides).

A snow ladybug taking a walk among galls on the surface of a poplar leaf.

TRIBE COCCIDULINI

Snow Ladybug

COCCIDULA LEPIDA LeConte *("KOKK-sidd-YOU-lah LEPP-pih-dah")*

Coccidula like marshes, their larvae do too,
In Russia it's "mire," but that sounds like goo!

THE NAME
Cocci refers to the berry-red colour typical of many ladybugs, and *Coccidula* is probably just a diminutive or derivative form of a general name for ladybugs. *Lepida* means pretty, neat or graceful. I call this the snow ladybug because I have found it more than once on warm days in March, crawling on snow near its marshy homes.

IDENTIFICATION
2.8–3.5 mm. An elongate, yellowish ladybug with a dark head, pale pronotum, and three broad dark stripes extending two-thirds of the way down the sides and the middle of the wing covers. The midline stripe is expanded into a spot near its tip. Compared with most other ladybugs, the snow ladybug looks more like a typical beetle, and they also have fewer and larger lenses in their compound eyes, and "open coxal cavities" where the legs join the body.

NOTES
This is a species of marshes and wet areas, and there it shares its home with such ladybugs as the marsh and the thirteen-spotted. They may be scale insect feeders, but no one knows for sure. The largest concentration of these beetles I have seen was in a sedge meadow near Lessard Lake, possibly feeding on leafhoppers. The somewhat hunch-backed larvae were there too. In Russia, the related species *C. rufa* is found in meadows and mire, and there are two *Coccidula* species (*C. rufa* and *C. scutellata*) in Britain that are both apparently much easier to find than ours is here. The snow ladybug is found all the way across southern Canada and the northern United States.

Subfamily Coccinellinae

This subfamily contains all of the so-called typical ladybugs. They are generally oval in shape, reasonably large (4 mm or more in length), and patterned in red or orange and black. There are, however, many exceptions. Some members of this group are primarily brown, and some are primarily black. Others are elongate in shape, and striped instead of spotted. But for the most part, members of this subfamily feed on aphids or related sucking bugs. One obvious exception is the wee tiny ladybug, which eats mold spores. In some books, the wee tiny ladybug is placed in a different tribe from the other members of the subfamily, but I favour an arrangement where all of our local species are members of a single tribe, the Coccinellini.

A marsh ladybug, typical of wet reedy places.

TRIBE COCCINELLINI

Marsh Ladybug

ANISOSTICTA BITRIANGULARIS (Say)
("ann-ICE-oh-STICK-tah BYE-try-an-gew-LARE-iss")

Where are the triangles? Why are there two?
Say didn't say, so what's there to do?

THE NAME
Aniso means uneven or unequal, and *sticta* refers to a row of things, in this case most likely spots on the wing covers. *Bitriangularis* means with two triangles, but I am not at all sure how this applies in this instance.

IDENTIFICATION
3.0–4.0 mm. A pale, yellowish elongate ladybug, with a wavy band near the outer edge of each wing cover, and a wavy line where the wing covers meet. These dark markings are broken into spots on some individuals.

NOTES
This is another species of marshes and sedges, and I have found it most abundantly in sedges at both Lessard Lake and near Opal. At Lessard Lake they were quite low down, and we found them by digging down into the sedge clumps by hand and by shaking the dead sedge into a beating net. At Opal, on a warmer day, they were higher up and we simply swept nets through the sedges as we walked through the marsh. Late in the summer at Opal, they seem to move up into nearby grasses, a few metres from the marsh itself. There were larvae at both locations as well, and one was right down in the dark stinky muck, in the company of very pale leafhoppers, which may have served as food. Robert Gordon says they are reputed to be aphid feeders, but he hadn't encountered any real evidence for this. Robert Gordon shows the related species *A. borealis* in extreme northern Alberta, but there are no confirmed records of *A. borealis* from Alberta, and this part of his map was based on speculation.

Episcopalian Ladybug

MACRONAEMIA EPISCOPALIS (Kirby)
("MACK-row-NEEM-ee-ah eh-PISS-koe-PAY-liss")

Episcopalian, bishop on high,
Deep in the sedges, looking up to a fly.

THE NAME
There is another genus of ladybugs named *Naemia*, so *Macronaemia* means greater *Naemia* (although some *Naemia* are larger than *Macronaemia*). *Naemia*, in turn, refers to pastures or woodland meadows. The name Episcopalian ladybug is taken from the scientific name, but it doesn't refer to the Episcopal Church. *Episcopalis* means bishop or, more likely, supervisor, with the added connotation of surveying from a height, possibly referring to the fact that these beetles sometimes (but not often) inhabit trees and bushes.

IDENTIFICATION
3.5–4.0 mm. A distinctive, elongate, yellowish ladybug with three narrow black stripes on the wing covers, and two black patches on the pronotum. Because of its body shape and stripes, it is often mistaken for a leaf beetle in the family Chrysomelidae (especially *Prasocuris phellandri,* and *Hydrothassa* spp.), and vice versa.

NOTES
This is the third marsh-dwelling species we have looked at in a row, but like the thirteen-spot further on in the book, it also turns up in trees, shrubs, and cropland. It inhabits my two favourite sedge marshes at Lessard Lake and Opal, and on one occasion at Opal they were less common than the marsh ladybug by a ratio of 7:15. Mike Majerus cleverly pointed out to me that they have the perfect colouration for this habitat, with longitudinal stripes that echo the

The Larger Ladybugs of Alberta   109

striping of the grasses and sedges that grow there, especially when they are brown and dead. Many other marsh-dwelling creatures share this pattern, such as the ground beetle *Agonum nigrescens*, the long-jawed orb weaver spiders, and *Leucania* moths. The Episcopalian ladybug is another unconfirmed but supposed aphid predator, according to Robert Gordon. My own introduction to the Episcopalian ladybug came when I was a teen-aged beetle collector, and a fellow teen named Larry Orsak wrote to me from California to ask if I could find him a specimen—he even included a sketch of the beetle. Larry was the most prominent ladybug aficionado in the then-magnificent Teen International Entomology Group, and his articles on ladybugs in the TIEG Newsletter were an early inspiration to me. It took me decades to find the species, and by then I assumed that Larry had acquired a specimen elsewhere. I believe he is still involved in entomology, however, so perhaps he will read this and let me know one way or the other.

The larva of the Episcopalian ladybug in the bottom of my beating net at Lessard Lake.

Thirteen-spot Ladybug

HIPPODAMIA TREDECIMPUNCTATA (Linnaeus)
("HIPP-oh-DAME-ee-ah treh-DESS-im-punk-TAY-tah")

Thirteen spots and thirteen abodes,
Marshes, lawns, and the sides of roads.

THE NAME
Hippodamia was a female figure in Greek myth. *Tredecim* means thirteen, and *punctata* means a rounded point, referring to the black spots on the wing covers. The name thirteen-spot ladybug is in wide use, and I prefer it to "thirteen-spotted."

IDENTIFICATION
4.5–6.4 mm. A semi-elongate orange ladybug with thirteen black spots on the wing covers, including the central spot that straddles the two wing covers next to the pronotum. The pronotum itself has an unmarked black centre. Also notice that on each leg the femur is black, but the tibia and tarsus are orange.

NOTES
The thirteen-spot in natural habitats is a species that occurs in wet meadows, sedge marshes, or near water. However, we most often find it in such places as grain crops and lawns. Perhaps there is something about cultivated grasses (in the broad sense, including corn and wheat) that reminds the beetles of wet meadows—this is true for some other sorts of insects, including the cherry-faced meadowhawk dragonfly (*Sympetrum internum*) that frequently lays eggs in lawns. Thirteen-spot ladybugs do need warmth, however—B.D. Frazer and R.R. McGregor found that this species did not develop at all, or lay any eggs, at 12°C. At the corn maze near Lacombe, I also found pupae (after all, what else is there to see in a corn maze), indicating that the larvae as well as the adults use

the man-made habitat. In Europe, it is clear that in the north and west this is a species of marshes, while in the south and east they live in crops and in herbaceous areas as well as in marshes. In Russia, Victor Kuznetsov also reports this species from both marshes and cropland. Thirteen-spot ladybugs often appear at wash-up and behave like typical members of their genus when it comes to searching for food. Our subspecies is *H. tredecimpunctata tibialis*, and it is found over most of Canada and the northern half of the United States.

A sight I have known since childhood—a thirteen-spot ladybug in the grass on the front lawn.

American Ladybug

HIPPODAMIA AMERICANA Crotch
("HIPP-oh-DAME-ee-ah ah-MERR-ih-CANN-ah")

"A remarkable beetle," thought the good Dr. Crotch,
Let's hope that its peatlands aren't burned to make scotch.

THE NAME
Americana refers to America, presumably in the broad sense. I have coined the name American ladybug here.

IDENTIFICATION
4.4–5.1 mm. American ladybugs have the typical semi-elongate shape of a *Hippodamia*, but unlike most members of the genus the dark markings on their yellow or yellow-orange wing covers form a central cross-shaped black mark and long two-pronged bands near the sides. The side bands may be broken into spots, and the pronotum has an unmarked black centre.

NOTES
This is a rare species that is apparently transcontinental in Canada. It is also likely to be a beetle of peatlands, although no one really knows for sure. John and Bert Carr found two dozen or so in a boggy area called Stauffer Lake near the town of Stauffer, and this is the basis for my own notion of its true habitat. It is also impressive how similar this species' markings are to two other species in the same habitat—the waterside ladybug and the marsh ladybug. American ladybugs have also turned up in such places as Adams Lookout in the Willmore Wilderness Area, Fidler-Greywillow Wildland Provincial Park, and in a variety of places that all strike me as having some black spruce or tamarack trees in the vicinity. I suspect that this is a species of cool places, and that we will eventually come to understand its habitat associations and thereby find it more easily. Vanessa Block found the Fidler-Greywillow beetle in 2001, so there is no reason to think that the species is no longer here.

Waterside Ladybug

HIPPODAMIA FALCIGERA Crotch *("HIPP-oh-DAME-ee-ah fall-SIDGE-err-ah")*

False igera? True igera? Or something in between,
You'll find the sickle-bearer in the reeds beside the stream.

THE NAME
I have coined the name waterside ladybug, since it seems clear that this is the habitat for the species. *Falcigera* means sickle-bearer, and may refer to the male genitalia, the "sipho" of which is indeed long, curved, and pointy.

IDENTIFICATION
5.0–6.0 mm. A good-sized, yellowish, semi-elongate ladybug with three black stripes on the wing covers; the outer two are wavy near the tip. The pronotum has an unmarked black centre. This is an easy species to recognize, although it is sometimes mistaken for a leaf beetle.

NOTES
The waterside ladybug is a species of western Canada and the adjacent United States. This is the third of the pale-coloured, striped, wetland ladybugs, along with the marsh and the American. I have only found it once, along the Prentice Creek in the patterned fen near Crimson Lake in the foothills. Records continue to trickle in for this species, however, and in each instance the beetles are found in waterside vegetation, usually along small streams (although we may not have enough records to say that this is always the case). It has shown up twice in collections made at the Environmental Sciences Centre of the University of Calgary, at Barrier Lake on the Kananaskis Highway (Highway 40), and at least one of these specimens was labeled "*Carex* marsh," *Carex* being the genus name for sedges. It also appears sparingly at wash-ups. There are some old records for Edmonton, but in a lifetime of beetling in Edmonton I have never encountered this beetle here.

Parenthesis Ladybug

HIPPODAMIA PARENTHESIS (Say) *("HIPP-oh-DAME-ee-ah parr-ENNTH-ess-iss")*

The parenthesis ladybug (one of the prettiest),
Is, parenthetically, also the wittiest.

THE NAME
The "parentheses" on this ladybug are obvious in the black markings on the wing covers of most individuals. The English name is in widespread use.

IDENTIFICATION
3.8–5.6 mm. The wing covers have parenthesis-shaped black marks near the tips; these parentheses may be expanded to form two dark blotches, or they may be reduced to four spots. Also notice the black shoulder spots, and a central black blotch that is roughly triangular or bell-shaped. The black markings on the pronotum are vaguely M-shaped. The ground colour varies from yellow-orange to dark orange and usually appears somewhat two-toned, blending from a lighter to a darker shade.

NOTES
Parenthesis ladybugs are widespread from the Yukon across Alberta and the southern portions of the provinces to the east, as well as the northeastern United States, the southwest, and parts of western Mexico. This is a common species, and it is especially easy to find at wash-up. They also appear in lawns, gardens, alfalfa fields, crops, and almost any other situation where there is low-growing vegetation. They are found on the prairies, in the forested areas of the province, and in the high country as well. Get to know the parenthesis well, since the less commonly encountered expurgate, sinuate, and boulder ladybugs

are similar. Over much of the province, this and the thirteen-spot are the two most common members of the genus *Hippodamia*. This is also one of the few species in our fauna that has been recorded coming to lights at night, although always in small numbers.

The hind end of a parenthesis ladybug, clearly without the black elytral tips that mark the expurgate ladybug.

Expurgate Ladybug

HIPPODAMIA EXPURGATA Casey *("HIPP-oh-DAME-ee-ah EX-purr-GATE-ah")*

There was once an expurgated insect,
That was soon, with a number 2 pin, decked.

THE NAME
Expurgata means deeply cleaned, perhaps in reference to the sparse markings of many individuals. I like the sound of the word, so I have coined "expurgate ladybug" as the English name.

IDENTIFICATION
3.5–5.0 mm. A lot like the parenthesis ladybug but with at least the tips of the wing covers forming a black spot where they meet (which may be extremely small on individuals with reduced black markings overall). The parenthesis ladybug always has orange wing cover tips and usually has more extensive black markings and a darker orange ground colour.

NOTES
This is a reasonably common species of ladybug on the prairies and a regular but infrequent species in the foothills and the eastern slopes of the Rocky Mountains. Joe Belicek called it *H. apicalis* (a separate species from west of the Rockies that can only be distinguished by dissection) and reported it on sagebrush infested with *Bryobia* mites. Ted Pike has also found huge numbers of this beetle by sweeping a net through sage on the prairies. I tried this, succeeded, and when I did it was clear that the beetles were feeding on aphids. I have also seen the expurgate ladybug at wash-up on Barrier Lake along the Kananaskis Highway and at the Ghost Dam west of Cochrane. In Alberta, this species has sparse markings, but elsewhere in its range the markings can be quite extensive and run-together. I personally mistook these beetles for parenthesis ladybugs for many years before realizing how easy it is to identify them, especially with a hand lens.

Five-spot Ladybug

HIPPODAMIA QUINQUESIGNATA (Kirby)
("HIPP-oh-DAME-ee-ah KWINN-kwess-igg-NATE-ah")

Quinquesignata, oh what is the trouble?
You are so hard to tell from your glacial double!

THE NAME
Quinque means five, and *signata* means mark or sign—logically the five-spot ladybug.

IDENTIFICATION
4.0–7.0 mm. This is the first of four similar species, each relatively large for the genus and with variable markings. In the prairies and the aspen parkland, individuals with white converging marks on the pronotum are probably five-spots, but those without are impossible to tell from glacial ladybugs without dissection of the males. In the southern Rockies, check potential five-spots carefully against the more-or-less distinctive Colonel Casey's ladybug, and the yellowish sorrowful ladybug. In the northern Rockies and the boreal forest, you are more-or-less justified in calling them all five-spots.

NOTES
In general, when you cannot identify a member of this group of similar species, it is best to consider it a five-spot by default, or to resist giving it a name. This is probably the most abundant of the large, difficult *Hippodamia*, even on the prairies, based on Robert Gordon's maps and my own admittedly limited dissections of a few dozen specimens. This species apparently made up the bulk of the ladybugs found by A.M. Harper and C.E. Lilly in southern Alberta. One aggregation, under bearberry plants, contained about 3,000–5,000 individuals. Another, under a juniper bush, had about 2,000 ladybugs in it.

Glacial Ladybug

HIPPODAMIA GLACIALIS (Fabricius) *("HIPP-oh-DAME-ee-ah GLAY-see-ALL-iss")*

Up in the mountain tops, icy and old,
You won't find the glacial, 'cause it's too stinkin' cold!

THE NAME
Glacialis means frozen, probably in reference to the northern range of the species. The English name is an obvious choice.

IDENTIFICATION
5.0–7.0 mm. This species is extremely similar to some colour morphs of the five-spot ladybug. In fact, females are impossible to distinguish, and males are identifiable only by dissection of their genitalia. Fortunately, the glacial ladybug is found only on the prairies and in the aspen parklands in Alberta, and at least some five-spots have white lines on the pronotum, which the glacial lacks (although some glacials have tiny white spots in their place).

NOTES
Our subspecies is *H. g. lecontei* Mulsant, and this is the race of the glacial ladybug that is most difficult to separate from the five-spot. Their similarity is probably the result of convergent evolution, in which the same forces that have selected for the colour pattern of one species are also acting on the other, since they live in the same places and do the same things. This sort of geographic co-variation is reasonably common in insects. So are species that are impossible to tell apart without reference to their genitalia (for example, the boreal and northern bluets, which I discuss in my book on Alberta damselflies). Some naturalists hold the belief that there must be some way to tell these sorts of species pairs apart, and that the short-sighted specialists have missed the obvious. These folks are usually birdwatchers, however, and there are many

physically identical pairs of bird species that have different songs, something ladybugs lack. I must add, however, that I was pleasantly surprised by the process of learning the appropriate dissections, and I was amazed at how well preserved the insides of specimens were, even if they were originally collected in the 1920s. I hope you realize, however, that for the most part I did it so you won't have to.

Am I 100% sure this is a glacial ladybug and not a five-spot? No.

Sorrowful Ladybug

HIPPODAMIA MOESTA LeConte *("HIPP-oh-DAME-ee-ah MEESS-tah")*

The sorrowful ladybug from the southwest,
Is actually, happily, one of the best.

THE NAME
Moesta means sorrowful, and this is the reason for the newly coined English name as well.

IDENTIFICATION
6.0–7.5 mm. Another large *Hippodamia*, with three large dark blotches on each wing cover. It is best told from the five-spot ladybug by its yellow or yellow-orange ground colour, rather than the typical medium or dark orange of most other *Hippodamia*. Some five-spots have exactly the pattern of dark markings you would expect on the sorrowful.

NOTES
Our subspecies is *H. m. bowditchi*, a beetle that is also found in British Columbia, Idaho, Montana, and Colorado. Robert Gordon shows it very close to the southwest corner of Alberta, but not in Alberta itself. In fact, I have not seen a dissected male specimen of this species from the province, and the only females I have seen are too orange to be certain of their identity. Kim Pearson brought me two specimens once from north of Waterton, and they both looked exactly like rather orange-ish sorrowful ladybugs, but the only male among them proved to be a five-spot when I dissected it. For the moment, I think this should be treated as a hypothetical but likely species in Alberta.

Colonel Casey's Ladybug

HIPPODAMIA CASEYI Johnson *("HIPP-oh-DAME-ee-ah KAY-see-eye")*

A ladybug named by the notorious Colonel,
Something to look for in Waterton vernal.

THE NAME
Caseyi honours Colonel Thomas Lincoln Casey, a prolific early describer of North American beetles.

IDENTIFICATION
4.8–6.7 mm. Very much like the five-spot but characteristically with the central spot at the base of the wing covers elongate or incorporated into a transverse band. Pronotum with the lateral extensions of the dark pronotal spot dividing the light border and with or without converging pale dashes in the dark centre. The colouration of this species is not as variable as the five-spot, and most individuals will more-or-less match the illustration.

NOTES
I have dissected one male specimen that was collected by F.S. Carr in Waterton on July 17, 1931. This is, as far as I am aware, the only record of a dissected male example of this species from Alberta, and it is worth noting that Bob Gordon did not list this species from the province. The Waterton specimen is a typical example of the species. This is not to say that I disbelieve the records published by Joe Belicek, or by A.M. Harper and C.E. Lilly in overwintering aggregations; however, it was not apparent from the writings of these authors whether or not they had performed dissections, and this may explain the absence of this species from Bob Gordon's very important work, since Bob understands well the need for such things. It will be interesting to chart the distribution of this species in the province now that we are better able to recognize it.

Convergent Ladybug

HIPPODAMIA CONVERGENS Guérin-Méneville
("HIPP-oh-DAME-ee-ah KONN-verr-JENZ")

That ladybug helper the gardener befriended,
Flew away leaving his plants undefended.

THE NAME
Convergens refers to the converging white lines on the pronotum. The name convergent ladybug is in widespread use.

IDENTIFICATION
4.2–7.3 mm. The pronotum has converging pale dashes in the dark centre, but this feature is shared by a number of other species. The wing covers usually have thirteen smallish spots, although the spots may be reduced on some individuals. Other species with the converging spots almost always have larger and fewer spots on the wing covers, or are noticeably smaller in body size.

NOTES
This is the ladybug that is collected in hilltop overwintering aggregations and sold to gardeners, especially in the United States. It ranges across much of southern Canada, although unevenly, and south to South America. I have found it sparingly but consistently in the southern half of the province, and it is a reasonably common species. A very similar species, the variegated ladybug (*Hippodamia variegata*) has been introduced in North America for Russian wheat aphid control (*Diuraphis noxia*). The variegated looks like a convergent ladybug but has the pronotal margin raised along the base and white, not black, bases to the front legs. It should be watched for in Alberta.

Boulder Ladybug

HIPPODAMIA OREGONENSIS Crotch
("HIPP-oh-DAME-ee-ah ORR-egg-onn-ENN-siss")

Oregonensis, from the land of snake fences,
On top of the mountain but never pretentious.

THE NAME
Oregonenis means "from Oregon," although the species is more widespread than that. Hence, I coin the name boulder ladybug in reference to its habitat.

IDENTIFICATION
4.0–5.0 mm. A small *Hippodamia*, the size of the parenthesis ladybug but with a colour pattern more like that of the sorrowful ladybug.

NOTES
Joe Belicek calls it an alpine meadow species that is often found on rocks, pupates on rocks, and aggregates in the winter, at least in British Columbia. The association of this species with alpine boulders is very much like that of the high-country ladybug. When Joe Belicek and Bob Gordon wrote their papers, the species was known in Alberta only from Banff National Park. There are, however, specimens of boulder ladybugs in the E.H. Strickland Entomological Museum from the Bald Hills in Jasper National Park, Pocaterra Creek in the Kananaskis Valley, as well as a spot near the headwaters of the North Saskatchewan River in Banff National Park. Greg Pohl and David Langor also found three of these beetles under rocks on top of Plateau Mountain.

Sinuate Ladybug

HIPPODAMIA SINUATA Mulsant *("HIPP-oh-DAME-ee-ah SYNE-you-ATE-ah")*

This ladybug is a sinuata,
But a sign you ate a what?

THE NAME
Sinuata means curved, or sinuous, and probably refers to some aspect of the wing cover markings (or perhaps the genitalia). I have therefore coined the English name "sinuate ladybug."

IDENTIFICATION
A small *Hippodamia* (the size and shape of the parenthesis ladybug) with convergent pale dashes on the dark center of the pronotum. Wing covers with a central dark mark near the top that is narrower at the rear than that of the parenthesis ladybug, and three, two or no black blotches on each wing cover. Sparsely marked, and a bit like a smaller version of the convergent ladybug.

NOTES
This is a western species with two subspecies, one of which is coastal, the other (ours) found over much of western Canada and the United States away from the coast. Our subspecies is *H. sinuata crotchi*, named for the famous Cambridge entomologist George Robert Crotch, a man with a particular love of ladybugs (and a close colleague of Hermann Hagen, whose name appears often in my book on damselflies). So far I have found sinuate ladybugs mainly at wash-up on lakes, so I am not really sure what their true habitat is. My only *in situ* encounter came through sweeping the salt-tolerant vegetation at the edge of a small salt lake north of Medicine Hat, but I doubt that sinuate ladybugs specialize in highly salty habitats. They are too widespread and too regularly encountered for that to be the case.

Flying Saucer Ladybug

ANATIS RATHVONI (LeConte) *("ah-NATT-iss rath-VONE-eye")*

Alberta is home to Anatis rathvoni,
Or was it a stray, and I'm spouting baloney?

THE NAME
Anatis means duck, perhaps because these beetles are often found on beaches at wash-up, but Willis S. Blatchley wrote that it comes from the Greek and means "harmless." Selina Stewart, my classicist friend, felt that it was vaguely possible that Blatchley was right but that there are easier ways to say "harmless," none of which are spelled exactly like the word for "duck." *Rathvoni* refers to Simon Snyder Rathvon, a Philadelphia entomologist who worked with scale insects.

IDENTIFICATION
7.5–10.2 mm. A very large and very round ladybug, with a 16 spotted pattern on the wing covers, and flanges off the sides of the wing covers. The pronotum has two white dots at the base.

NOTES
A Pacific Northwest species. Joe Belicek suggested that the one and only Alberta record, from wash-up at Waterton Lake, was a windblown stray from farther west, and I agree with him, especially after Mira Snyder spent an entire summer searching for ladybugs in Waterton and failed to find it. W.Y. Watson, who reviewed this group of ladybugs in the Canadian Entomologist in 1976, says all of our *Anatis* species prefer to feed in coniferous trees.

Large Orange Ladybug

ANATIS LECONTEI Casey *("ah-NATT-iss leh-KONT-eye")*

A ladybug named for ol' John Le Conte,
Large, orange and rounded with beauty to flaunt!

THE NAME
Lecontei honours John L. Le Conte, a very important early American entomologist. Large orange ladybug is a newly coined English name, which echoes the English name for a butterfly of the southern United States, *Phoebis agarithe*, the large orange sulphur.

IDENTIFICATION
7.8–10.5 mm. A large rounded ladybug with the wing covers plain orange and with a dark outer flange-like border.

NOTES
Found from southern Alberta south almost to Mexico, in a reasonably narrow band. Most of the Alberta records come from the Crowsnest Pass, but the species has been recorded as far north as the Ghost Dam (on the Bow River west of Cochrane) and as far east as Claresholm (where it was found on willows, not coniferous trees). This is still a poorly known ladybug in Alberta, known from less than 20 records, but it is also clearly established here, unlike the flying saucer ladybug. Our most recent record is from 1980, so any new sightings are of interest.

American Eyespot Ladybug

ANATIS MALI (Say) *("ah-NATT-iss MAHL-eye")*

If ever you find yourself down a dark alley,
Hope that you meet a few Anatis mali!

THE NAME
Mali refers to apples and was probably intended to indicate that the species prefers to live on apple trees. The name American eyespot ladybug is intended to distinguish this species from other eye-spotted members of its genus, especially *A. ocellata*, a species of Europe and Asia.

IDENTIFICATION
7.3–10.0 mm. Easy to recognize because of its large size, rounded shape, and the light halos surrounding the 18 or so dark spots on the wing covers.

NOTES
This is our most spectacular ladybug, and what a wonderful coincidence it is that our largest ladybug is also our most fancily coloured. The American eyespot ladybug is actually a transcontinental species ranging from southern Alaska to the northeastern United States. Joe Belicek believed that the original name *A. mali* needed replacing and proposed the replacement name *A. borealis*, but others have shown that "*A. mali*" should be retained. This species is clearly a forest dweller. In a jack pine forest near Devon, west of Edmonton, Chris Schmidt and Gary Anweiler were out enticing moths to a light on a warm evening in early November 2005, and enough ladybugs came to the light that they made a note to let me know. Among the ladybugs that came to the sheet were two or three American eyespot ladybugs, coaxed out of their hibernation sites by the ultraviolet glow.

Streaked Ladybug

MYZIA PULLATA (Say) *("mye-ZEE-ah poo-LATT-ah")*

My zia, your zia, whose zia is zia?
That zia, this zia, frankly I have no idea!

THE NAME
There is an aphid genus *Myzus*, that includes the tobacco aphid (*M. nicotianae*) and the peach-potato aphid (*M. persicae*). *Myzus* refers to sucking, which is what aphids do, and *Myzia* probably refers to the importance of these aphids to at least some members of this ladybug genus. *Pullata* means clothed in a soiled garments, probably because of the unusual streaked appearance of this beetle. The English name "streaked ladybug" is therefore coined here.

IDENTIFICATION
6.5–8.0 mm. A large ladybug, usually grayish in ground colour, and with a streaked pattern on the wing covers (reduced on some individuals). Look for the dark spot within the pale border of the pronotum.

NOTES
The genus *Myzia* is closely related to *Anatis* and, historically, some specialists have tried to unite the two under one name. Members of both groups seem to prefer forests as their habitats. Mike Majerus tells me that in England the striped ladybug (*Myzia oblongoguttata*) breed on the pines but can fatten up for the winter on aphids on a variety of trees. When he was here and we were exploring for ladybugs near Opal, we found both adults and larvae of the streaked ladybug on a lone jack pine tree, in the company of some very large aphids. The larvae are blackish and long legged. Mike also found them (he seems to have the knack) on a lone spruce tree near Crimson Lake, among many wooly aphids, and on a birch tree in the parking lot of the Callingwood Shopping Centre in Edmonton. Clearly, the trick is to search more lone trees.

Subvittate Ladybug

MYZIA SUBVITTATA (Mulsant) *("mye-ZEE-ah SUBB-vitt-ATE-ah")*

Subvittata, *almost-banded,*
The name is rather even-handed.

THE NAME
Subvittata means almost banded. Since the word subvittate is pleasant to me, I have used it in the English name for the species.

IDENTIFICATION
5.7–8.0 mm. Another large ladybug with a streaked pattern on the wing covers. The white side borders of the pronotum do not have black spots within them. Subvittae ladybugs are red in ground colour when they are mature but more brownish when they are young.

NOTES
This is a very fine looking ladybug, especially when it assumes its red ground colour as a full adult. I have found it at Waterton, in the wash-up of Waterton Lake. In general, it is a species of the west coast of the United States, but here in Alberta it is found in the southwest corner, and especially along the Crowsnest Pass, as well as in the Cypress Hills as an isolated population. This distribution pattern suggests strongly that the subvittate ladybug is tightly tied to coniferous forests, and there are many other insects whose range includes the southern Rockies in Alberta as well as the Cypress Hills. There is one specimen, in the E.H. Strickland Entomological Museum, from a white spruce tree in Maple Creek, Saskatchewan, near but not in the Cypress Hills. As well, a single specimen of *M. interrupta* was reported in Joe Belicek's paper as an accidental occurrence in Medicine Hat. This species is unlikely to occur here as a breeding population but should be watched for just in case. It is paler overall than the subvittate ladybug and has no dark markings on the head.

Polkadot Ladybug

CALVIA QUATUORDECIMGUTTATA (Linnaeus)
("KALL-vee-ah kwat-YOU-orr-DESS-imm-goot-ATE-ah")

Polkadot Calvia, *pretty in pink,*
Sipping on beer with molasses to drink.

THE NAME
Calvia means hairless or bald. *Quatuordecim* means fourteen, and *guttata* refers to drop or teardrop-shaped marks. Polkadot ladybug is a new name that nicely describes at least the pink morph of this species in Alberta.

IDENTIFICATION
4.0–5.5 mm. Two colour forms of this medium-sized, rounded ladybug are found here. One is pink with round black spots, the other is black with yellow spots. Both are easy to recognize. A pale form is rare.

NOTES
This is a widespread ladybug that also occurs in Europe and parts of Asia but as a different colour morph. Mike Majerus, when he was here in Alberta, found it almost impossible to believe that ours are the same species as those in Britain; however, he has also been doing some breeding experiments using both in his lab, and they interbreed well, allowing Mike to work out some of the genetics of colour pattern in this species. My experience suggests that this is a species of deciduous trees, and a number of authors have suggested that this species prefers to feed on psyllid planthoppers but also takes aphids. At Lac La Biche, I have seen this beetle come to a bowl of carrot and oatmeal babyfood on a picnic table in July and once, on August 27, I found a polkadot ladybug on two-day-old moth bait (beer, sugar, rum, and molasses) at Gull Lake—I assume it was looking for secondary food sources to fatten itself up before winter time.

Two-spot Ladybug

ADALIA BIPUNCTATA (Linnaeus) *("ah-DAY-lee-ah BYE-punk-TATE-ah")*

*Every one different, the two-spotted ladies,
Happy in forests protected by* Aedes.

THE NAME

The well-established English name "two-spot ladybug" is unfortunate, given the many colour forms that are not two-spotted. *Adalia* appears to mean uncertain, but even that is uncertain. Blatchley calls it "an invented name," but an anonymous reviewer of this book pointed out that *Adalia* is also is a biblical Hebrew name. *Bipunctata*, however, clearly means two rounded points.

IDENTIFICATION

3.5–5.5 mm. This is a medium to smallish ladybug and a very familiar beetle. The most common form is red with two black spots on the wing covers. These can be doubled to form a row of four smaller spots or missing altogether. There is also a two-banded form and a melanic form with pale shoulders. On the prairies, they often have 10 small spots in the same pattern as the Halloween ladybug. To recognize some of the variations, look carefully at the pronotum with its M-shaped central mark and the black dots in the pale sides.

NOTES

The two-spot ladybug is the species most often encountered hibernating in buildings in Alberta, and it also hibernates on trees in warmer places than our province. These hibernating groups are small, usually comprised of 20 beetles or less. Frazer and McGregor found that this species had the lowest developmental threshold temperature for egg development (8.4 °C) of any of the species they studied, giving the two-spot an early start in the springtime. The two-spot is an arboreal ladybug, and I have found it in good numbers on caragana (which often becomes covered in aphids), mountain ash, birch (especially weeping birch), and wolf willow. Ivo Hodek and Alois Honěk say that the larva of this species has a very well developed adhesive organ on its tenth abdominal segment, and that tree-dwelling ladybugs in general are more likely to hold onto the vegetation when disturbed, compared with herb dwellers. This is a variable species, but originally it was thought that the North American populations of the two-spot were not as variable in pattern as the Eurasian populations, and that this might suggest a "founder effect," where our populations descended from a limited genetic sample from the Old World. In other words, they may have been introduced. At Iowa State University, the genetic variability of the two-spot was examined, and good evidence emerged that the two-spot did not arrive in North America in historical times. In Alberta, this is a wonderfully variable species, so much so that the founder effect hypothesis seems ridiculous. Many of the uncommon forms of the two-spot were once considered a separate species "*Adalia frigida.*" It has also been suggested that the two-spot is a mimic, and that the two-spotted form mimics the more common seven-spot ladybug in Europe. Here, it makes sense that the banded morph is common where the transverse ladybug was once the most abundant bad-tasting ladybug (although the banded two-spot looks more like the three-banded ladybug, at least to people). Will this change now that the transverse is rare? This could be a very interesting area for future research.

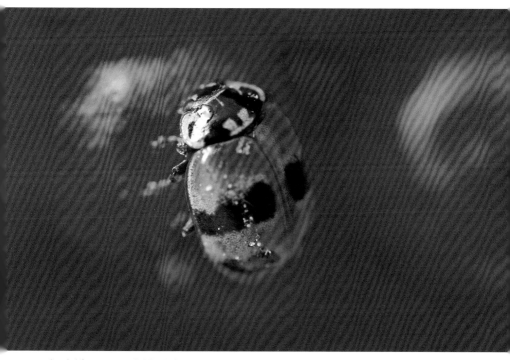

A banded-form two-spot ladybug hiding out among some mountain ash berries.

Three-banded Ladybug

COCCINELLA TRIFASCIATA Mulsant *("KOKK-sinn-ELL-ah TRY-fass-see-ATE-ah")*

*I dreamed that I found a nice trifasciata,
Stuck to the side of my sister's Miyata.*

THE NAME
Coccinella refers to a scarlet or berry-red colour and the association of that colour with the Virgin Mary, who is the "lady" in "ladybug." *Trifasciata* means three bands. The name three-banded ladybug is in widespread use.

IDENTIFICATION
4.0–5.0 mm. A rounded ladybug with a sunburst ground colour, a dark band across the shoulders, and two broken dark bands to the rear of that. In this sense, it is much like the transverse ladybug, but generally smaller, and with the two broken bands much wider (and sometimes faintly outlined in yellow). The banded form of the two-spot ladybug is often misidentified as the three-banded, but the two-spot never possesses a shoulder band.

NOTES
This is a widespread species, which lives in Europe and Asia, as well as much of transcontinental North America. The widespread North American subspecies is *C. t. perplexa*, while in the Pacific Northwest you find *C. t. subversa*, a different looking ladybug with reduced markings on the wing covers. In Eurasia, they look much like our Alberta beetles. Three of our native *Coccinella*—the three-banded, transverse, and hieroglyphic—are also found in Eurasia alongside the seven-spot ladybug. This makes it easier to understand why the arrival of the seven-spot did not result in the extinction of its native close relatives, but rather a change in which species is most common. Is the three-banded ladybug a habitat specialist, as opposed to a generalist like the transverse once appeared to be, and the seven-spot is now?

Transverse Ladybug

COCCINELLA TRANSVERSOGUTTATA Falderman
("KOKK-sinn-ELL-ah tranz-VERSS-oh-goot-ATE-ah")

The rapid decline of transversoguttata,
Is apparent from all of our specimen data!

THE NAME
Transversoguttata refers to the transverse dark mark that extends across the top portion of both wing covers. *Guttata* also implies drops or teardrop shapes. The English name transverse ladybug is well established.

IDENTIFICATION
5.0–7.8 mm. A red-orange, rounded ladybug with a dark band across the shoulders and four dark blotches farther back on the wing covers. Transverse ladybugs are larger than three-banded ladybugs, and the four dark blotches are smaller on the transverse as well.

NOTES
Our subspecies is *C. t. richardsoni*. This was certainly the most abundant, widespread ladybug in Alberta up until the arrival of the seven-spot. Wendy Harrison and I demonstrated that the relative abundance (among the aphid-feeding ladybugs other than the seven-spot) of transverse ladybugs declined steeply during the decade after the seven-spot's arrival. As I mentioned in Chapter 4, this species is now easiest to find on the crests of partly vegetated sand dunes and on sparse vegetation on pediment slopes in badlands. Mike Majerus and I explored the possibility that root-feeding aphids were the prey in these habitats, but this seems not to be the case. Coincidentally, while pondering this problem on the Opal sand dunes (the same spot mentioned under the tamarack ladybug), I photographed a European skipper butterfly,

only to discover the image of a tiny grass-feeding aphid in the picture, right where we had been finding the ladybugs. Mike and I, along with Dave Lawrie (now widely known in Alberta entomological circles as "Physics Dave," since he is, by training, a physicist) also found five transverse ladybugs on seedlings of balsam poplar and willow in a recently burned black spruce bog north of Edmonton. This species' shift from an abundant habitat generalist to an uncommon habitat specialist is a rare and interesting phenomenon, and the transverse ladybug's newfound habitat associations deserve further study, much of which can be done by amateur naturalists.

It is still possible from time to time to find a transverse ladybug in their old haunts, lawns included.

Seven-spot Ladybug

COCCINELLA SEPTEMPUNCTATA Linnaeus
("KOKK-sinn-ELL-ah SEPP-temm-punk-TATE-ah")

Seven-spot, that awful blot,
It ain't as bad as we originally thought!

THE NAME
Septempunctata means seven rounded points, refering to the black wing cover spots. "Seven-spot ladybug" (or seven-spotted) is a well-established name.

IDENTIFICATION
6.5–7.8 mm. A red-orange rounded ladybug with seven black spots on the wing covers. Note that the nine-spot ladybug is similar but is paler in colour, and the line where the wing covers meet is black on the nine-spot as well.

NOTES
The story of this ladybug is now quite widely known. It was originally introduced to North America beginning in 1956 and became established in New Jersey in 1973. The spread of this species in the west was helped along by additional releases in response to the arrival of the Russian wheat aphid in the 1980s. It is now found throughout Alberta in most habitats, although the adults appear in places where we have never seen larvae, such as the alpine zone in the mountains. As well, this is a species of low vegetation and is not common in trees. In general, the seven-spot seems to have usurped the transverse ladybug's role as the most abundant habitat generalist. It is not the most abundant species everywhere, however, and some entomologists think

that it has peaked and is now becoming less abundant—a pattern common among introduced species. Vanessa Block found that, in the Canadian Shield in northeastern Alberta, seven-spots were only the third most abundant ladybug species after five-spots and two-spots. On the prairies, they are most often encountered on roadside "weeds," such as Canada thistle, sow thistle, and sunflowers. Seven-spot ladybugs are most visible in the fall when they gather to hibernate, always on the ground and under something, such as leaf litter or the edge of a building's foundation. Dobzhansky studied the life cycle of this species in the Ukraine (where it is probably similar to Alberta) and found that the first generation each year emerged and began to reproduce in early summer. The second generation emerged later in the summer but did not reproduce until after hibernation the following year. This pattern was shown to be determined genetically by alternation of generations, since a particularly late year in 1921 resulted in first-generation beetles emerging when second-generation beetles normally would and behaving like first-generation beetles despite this (i.e., they produced a late cohort of doomed larvae). I have seen pupae in July in Edmonton and in mid-September in Medicine Hat, as well as newly emerged beetles in early August at Gull Lake, suggesting that this species produces three generations in most of Alberta. In the Edmonton area, I have also found early September aggregations of these beetles on dry plants, apparently aestivating while food is scarce.

Older seven-spot ladybugs are dark red, while newly emerged adults are more orange in colour.

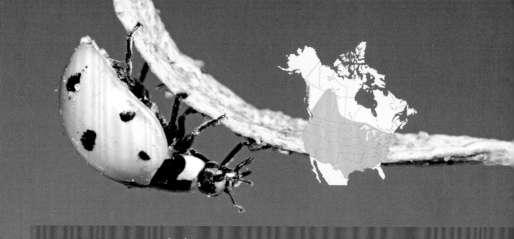

Nine-spot Ladybug

COCCINELLA NOVEMNOTATA Herbst
("KOKK-sinn-ELL-ah NO-vemm-note-ATE-ah")

Easterly nine-spots are no longer found,
But I know they still live here, so go look around!

THE NAME
Novem means nine, and *notata* means marks. The name nine-spot (or nine spotted) ladybug is in wide use.

IDENTIFICATION
4.7–7.0 mm. A pale orange, medium to large, rounded ladybug with nine black spots on the wing covers and a dark line where the wing covers meet. Likely to be confused only with the seven-spot, a brighter orange beetle without the dark line.

NOTES
Although this species has been recorded in Alberta as far north as Opal, I associate it most strongly with the prairies and the southern foothills. It is regularly found at wash-up at both the Ghost Dam and at Barrier Lake. As outlined in Chapter 4, this species has apparently disappeared from much of the eastern United States. Thus, its continued survival here in Alberta is a source of interest to ladybug biologists. Many of my ladybug colleagues have been skeptical that the nine-spot could survive here, but we continue to find them in small numbers. I hold out hope that we might find that this species is associated with a specific habitat here, similar to the transverse ladybug. I have found nine-spot ladybugs among scurf pea plants on the edges of sand dunes and blowouts at both the Purple Springs sand hills in Alberta and the Burstall dunes

in Saskatchewan. This may be their most predictable habitat right now. Of course, I may be entirely mistaken, and the nine-spot ladybug might go extinct in North America, as many suspect it will. Only time, and careful observation, will tell. R.D. McMullen studied the life cycle in this species and found that it went through two generations. Spring-generation adults went into diapause in the heat of summer, in response to increasing day length. Summer-generation adults went into winter diapause in response to decreasing day length. Only a day of about 16 hours seemed not to induce diapause (and presumably early spring day length didn't either).

Eating sticky, gooey aphids makes ladybugs sticky themselves, and they are often covered in lint that shows up annoyingly well in close-up photographs!

High-country Ladybug

COCCINELLA ALTA Brown *("KOKK-sinn-ELL-ah AWLT-ah")*

The high-country ladybug up near the snow,
Few people find it, but few people go.

THE NAME
Alta means high and refers to the high-altitude habitat of this species. Hence, I have coined "high-country ladybug" for this species.

IDENTIFICATION
4.8–5.3 mm. A rounded, red-orange ladybug with a diamond-shaped central spot between the shoulders, two large spots halfway down the wing covers, and two smaller spots near the tip. It closely resembles the tamarack ladybug, but where the wing covers meet on the high-country ladybug there is a dark line. The overall pattern is also much like that of five-spot and the sorrowful ladybugs, but these two lack the white front corners on the pronotum and are more elongate overall.

NOTES
This is one of our least common ladybugs, or perhaps one of our least frequently encountered ladybugs. We have records only from the mountains, from Banff down to Table Mountain near Beavermines Lake. When Bob Gordon published his work in 1985 there were only 10 localities known for this beetle anywhere in its known range (with a huge gap between Alberta and California). Gerry Hilchie found the Table Mountain beetle in 1995, providing evidence that the species is surviving since the arrival of the seven-spot.

    Mike Majerus and I also found a single high-country ladybug larva above timberline in the southern Highwood Pass, at 2300 metres elevation on a rock

The one and only larva of the high-country ladybug known to science.

(much like the classic habitat of the boulder ladybug), on August 13, 2000. The larva metamorphosed into an adult, thus confirming its identity. The larva of this species has never been described, so I think it is appropriate to do so here, based on photographs of the living larva and microscopic examination of the crumpled exuviae (shed skin), recovered from the remains of the pupa after the adult emerged. I was amazed to discover that one has to learn a whole new vocabulary to formally describe ladybug larvae (based on the classic work of J. Howard Gage in 1920). For example, a small bump with a single "hair" (seta) on it is a chalaza, whereas a larger bump with many setae is a parascolus (not to mention "strumae"). With apologies for that and the other jargon necessary for this process, here is the description.

The length of the larva was not measured in life but was probably greater than 7 mm and less than 10 mm. The characters used in Gordon and Vandenberg's key (in which this species keys to *C. undecimpunctata* because the tarsal claws lack a basal tooth), as well as those clearly visible in the photographs, are described as follows. Other characters were not examined, because the material at hand is not adequate for an exhaustive description. Dorsal surface of body without numerous small, scattered chalazae. Head black basally (the black area surrounding the ocelli, with its anterior border extending transversely across the head), dark yellowish distally, with pale but black-tipped antennae and maxillary palpae. Ground colour of body dark, bluish gray, with orange areas on the anterior and posterior pronotal angles, and around the dorsolateral and lateral parascoli of abdominal segments I and IV. Legs black, tarsal claws with basal area lacking a subquadrate tooth. Dorsolateral pale areas of abdominal segments I and IV narrowly but distinctly separated from dorsal parascoli by a dark area. Dorsal chalazae apparently with black apical setae, as judged from the photographs. Pronotal plate with chalazal bases shorter than antennae. Pronotal and metanotal chalazae with apical setae slender and much longer (four times or more) than chalazal bases. Metanotal plate with five large chalazae on outer margin. Height of dorsal parascoli on abdomen greater that half the distance between the two dorsal parascoli. Tergum of segment IX with approximately 60 small chalazae. The larval and pupal exuviae are deposited in the E.H. Strickland Entomological Museum. The pupal exuvia conists of numerous small pieces, and since the pupa was not photographed alive, no pupal description is attempted here.

I should also tell you here that Mike was at first disgusted, then skeptical of my identification of the stain on his pants, and later quite proud to have slipped in a pile of grizzly bear scat while searching for this larva. For Englishmen, encounters with large dangerous animals are rare and something to brag about back home! Sadly, however, a hiker was badly mauled the week after our visit not far from where we were searching and probably by the same bear that left its mark on Mike.

Some of the alpine zone ladybugs show up from time to time at wash-up in lakes near the mountains, but so far the high-country ladybug has not been among them.

Tamarack Ladybug

COCCINELLA MONTICOLA Mulsant *("KOKK-sinn-ELL-ah monn-tih-KOE-lah")*

My text is now due, it's the first week of March,
And finally I grasp that they're probably on larch!

THE NAME
Mont means mountain and *cola* means inhabiting. Thus, "mountain ladybug" would be an appropriate name, but because the species is more widespread than that, I prefer to call it the tamarack ladybug for reasons that are outlined below.

IDENTIFICATION
5.2–7.0 mm. A red-orange, round ladybug with a central triangular dark spot between the shoulders, four prominent dark spots farther back on the wing covers, and in some individuals, a pair of tiny spots as well.

NOTES
Bob Gordon calls this species widespread but uncommonly seen and that certainly describes it's reputation in Alberta. To look at the distribution map, you might think you could find it anywhere, but this is easier said than done. However, there does seem to be a bit of a pattern in the data: some localities (Hondo and Hay River, for example) have produced numerous records from the same date, and in each such instance the beetles were on tamarack trees (= larch, *Larix laricina*), a peatland species of bogs and fens. I don't know about you, but I'm keen to get out the beating sheet during June, July, and August and look for these beetles on tamarack trees. If indeed this is their preferred habitat, it will be interesting to find out what they are feeding on and whether the seven-spot ladybug inhabits the same trees. It may not be as easy as that, however. In 1985, the Provincial Museum of Alberta surveyed many thousands of beetles

at the Wagner Natural Area, a fen, west of Edmonton, and did not find this species (although perhaps they didn't use appropriate search techniques for this species). My own experience with this species comes from one encounter, in the spring of 2006, when my wife and I found three individuals on a sand dune near Opal, where we were accustomed to finding transverse ladybugs. That day, the transverse were nowhere to be seen; however, a week later the tamarack ladybugs were gone, and a few transverse were back where we expected them.

With its legs and antennae tucked in, this tamarack ladybug would be difficult for an ant to harm.

Hieroglyphic Ladybug

COCCINELLA HIEROGLYPHICA Linnaeus
("KOKK-sinn-ELL-ah HYE-row-GLIFF-ick-ah")

Hieroglyphic, sounds terrific,
But its food habits seem to be very specific.

THE NAME
Hieroglyphica refers to the wing cover markings, which resemble ancient Egyptian symbol-based writing, at least a bit. The name hieroglyphic ladybug is well established.

IDENTIFICATION
3.7–4.7 mm. A medium-sized, red-orange, round ladybug with a thick somewhat W-shaped band across the shoulders, and a pair of large spots toward the rear of the wing covers.

NOTES
This is a ladybug of the north, found in North America as well as Europe and much of Asia. Mike Majerus points out that hieroglyphic ladybugs eat aphids in some parts of Europe and Asia, but that they eat leaf beetles in Lapland, and are impossible to rear on aphids in Britain. It is possible that our Alberta populations also require leaf beetles, but I really don't know for sure. Specimen records from other places suggest that this is a tree-dwelling species, and it has been found on poplars, spruce, willows, and tamarack. Near Bently it was found "with *Archips sp.* on aspen." *Archips* is a genus of leaf-rolling tortrix moths. In St. Albert the hieroglyphic ladybug was found "feeding on *G. decora*," presumably referring to what is now known as *Tricholochmaea decora* (formerly *Galerucella decora*), a common leaf beetle on willows and aspen poplar.

Polished Ladybug

CYCLONEDA POLITA Casey ("SYE-kloe-NEED-ah poh-LITE-ah")

A pure red ladybug, little and round,
That lives on the plants down close to the ground.

THE NAME
Cyclo means a circle (and this is a very round beetle), and *neda* is a reference to a nymph from Greek mythology, who nursed Zeus when he was a baby. *Polita* means polished, hence the English "polished ladybug."

IDENTIFICATION
3.5–6.2 mm. The polished ladybug is a red, round ladybug with no black spots. Its identification is best confirmed by the four-fingered black pattern on the pronotum. Be especially sure you don't confuse the polished ladybug and the unmarked morph of the two-spot.

NOTES
This is a western species that enters Alberta only in the southwest. Bob Gordon and Nat Vandenberg say the genus *Cycloneda* is closely related to *Coccinella*, and *Cycloneda* replaces *Coccinella* in South America and has similar habitat and prey preferences. It is certainly true that I have seen plenty of *Cycloneda* in southern Texas and no *Coccinella*. Most of the time when you find an entirely unmarked red ladybug, it will be this species. The two-spot and the Halloween ladybugs can also be unmarked, however, as can the California ladybug, *Coccinella californica* (this orange beetle appeared a few times in Alberta in the 1970s, and probably arrived on produce). These are more orange than the polished ladybug, and larger as well, but I know of no recent records of this species in the province.

Halloween Ladybug

HARMONIA AXYRIDRIS (Pallas) *("harr-MOE-nee-ah ax-ee-RIDD-riss")*

Harmonia *the mythical, the one we love to hate,*
Creeping up the continent to poop upon our gate…

THE NAME

Harmonia means a fitting together—possibly because the beetles crowd together at the overwintering sites. *Harmonia* was also a figure in Greek mythology, married to *Cadmus*, who was responsible for the alphabet (and perhaps the markings on these beetles reminded Étienne Mulsant of Greek letters when he coined the name). *Axyridris* is more difficult to interpret but probably comes from *achyros*, meaning chaff, husk, or bran, and *idris* meaning skilled—as if they were skilled in husking seeds. It's hard to say. The Canadian Nature Federation referred to this species as the southern ladybug, but the name (which makes little sense) never caught on. In the United States, where the species can be a problem in houses, it has been given the name Asian multicoloured ladybug. I dislike this name because there is nothing obviously "multicoloured" about the beetle, and I would rather avoid any resonance between nativism and racism and therefore do not wish to implicitly blame Asia for the lousy reputation of this fully established North American beetle. The British call it the harlequin ladybug, but I suggest we use another North American name, Halloween ladybug, since the beetles are indeed pumpkin orange and are most noticeable when they congregate in the fall.

IDENTIFICATION

A very round, medium-large ladybug, usually pumpkin orange with a 10-spotted pattern on the wing covers. This pattern is shared by some prairie two-spots, but the Halloween is larger and always has a bit of a transverse crumpled area near the tips of the wing covers. The Halloween also has an unmarked morph and a melanic morph, but I don't think either of these has appeared in Alberta thus far.

NOTES

I have already written quite a bit about this beetle elsewhere in the book, but it should be emphasized that it is not yet clearly established in Alberta—we have only a few records from around Olds; these may be beetles that have escaped from greenhouses, where they are sometimes used for pest control. This is a species of both trees and low vegetation—a real habitat generalist. The adults can live 3 years. One of the most interesting aspects of this beetle is its tendency to aggregate in buildings for the winter. Victor Kuznetsov says, "Synanthropic accumulations of lady beetles in the south of the Far East are less common than hypostatical. The former are recorded in dwelling places, in production buildings, forest huts, barns, apiaries and other man-made structures. The aggregations are particularly extensive in taiga areas where *H. axyridis* and *A.* [*Aiolocaria*] *hexaspilota* accumulate in this manner. Having reached the walls of the structures, the beetles gradually work their way into cracks and hibernate there. Sometimes huge numbers of beetles accumulate in houses. They penetrate everywhere, settle on people and annoy them with their slight pinching.

The rare black form of the Halloween ladybug, looking a lot like a twice-stabbed ladybug but with white on the pronotum.

The wedge-shaped maxillary palps typical of larger ladybugs, and the four-segmented tarsus, are both visible on this individual.

Ignorant villagers in the taiga mistake these lady beetles for the serious potato pest, *H.* [*Henosepilachna*] *vigintioctomaculata*, sweep them off the walls and burn them." Where the species is well established in the United States, people are no less happy about its winter gatherings. Eric Riddick and his colleagues, from the Agricultural Research Center in Beltsville, Maryland, found that adult beetles were repelled by both camphor and menthol, although they fell short of recommending that we all spray our houses with these chemicals. Presumably, their study is a step toward finding a way to prevent the problem. Christine Nalepa and her North Carolina colleagues found that Halloween ladybugs were strongly attracted to constrasting patterns of black and white stripes in the fall, so perhaps there is a lesson there, to avoid contrasty houses or siding. Louis Tedders and Paul Schaefer found that they were caught in white weevil traps (made from interlocking masonite panels) more often than in brown, gray, or black traps. I have heard it said that they aggregate on pale-coloured cliffs in China, and I have certainly seen this in southern Illinois as well.

Painted Ladybug

MULSANTINA PICTA (Randall) *("mull-SANT-ee-nah PICK-tah")*

The petite painted lady of Étienne Mulsant,
Some say she climbs trees, while some say she can't.

THE NAME
The name *Mulsantina* honours Étienne Mulsant, an entomologist who studied ladybugs in the 1800s, and *picta* means painted. I think the name "painted ladybug" is appropriate for this species, and it echoes the name of a familiar butterfly, *Vanessa cardui*, the painted lady.

IDENTIFICATION
3.3–5.3 mm. The painted is a small- to medium-sized, rather parallel-sided, brownish ladybug with a pattern of dark wavy lines of the wing covers that are connected across the midline. Easy to confuse only with the Hudsonian ladybug, it's closest relative.

NOTES
The *Mulsantina* ladybugs may well be tree-dwellers, or at least, that is the impression we get from the few records that are not from lake wash-up. This is a beetle that I have encountered quite a few times but never in what appears to be its habitat. It is a slow flier, and more than once, I have snatched one out of the air with my hand. Like the Hudsonian labybug, it has been recorded from the Cypress Hills as well as other forested areas in the province.

Hudsonian Ladybug

MULSANTINA HUDSONICA (Casey) ("mull-SANT-ee-nah hud-SONN-ih-kah")

A nice little beetle is the fine Mulsantina,
Perhaps Mulsant drank once at Rosa's Cantina.

THE NAME
Hudsonica refers to Hudson's Bay and was used often in the early days of taxonomy to refer to anything northern. Thus, the English name Hudsonian ladybug.

IDENTIFICATION
3.5–5.0 mm. A pale tan, parallel-sided ladybug with a pair of curvaceous markings running the length of the wing covers but not joined across the midline, as well as a pair of small spots toward the tip of the wing covers.

NOTES
From our specimen and observation records, it is clear that the Hudsonian ladybug is a tree species, without much concern for which tree. I have found it in all the places one finds polkadot, two-spot, and other arboreal ladybugs, as well as in black spruce forests. The Hudsonian ladybug is found in the forested regions of the province and also in the Cypress Hills. For a long time, I was curious to know if this beetle is bad tasting, since it is tan in colour instead of red or orange, and I'm now happy to report that it's reflex blood is mildly bitter but certainly among the least distasteful of the ladybugs I have tested.

Wee-tiny Ladybug

PSYLLOBORA VIGINTIMACULATA (Say)
("sye-loh-BORR-ah vig-INN-tee-MACK-you-LATE-ah")

A wee tiny ladybug feeding on mould,
Loves life in Alberta, no matter how cold.

THE NAME

Psyllobora refers to the fact that these ladybugs were thought to devour psyllid plant hoppers. They are now understood to be mildew feeders. *Vigintimaculata* means twenty-marked. Here in Alberta, I started calling these beetles "wee-tiny ladybugs" many years ago, and the name seems to have caught on among my friends at least.

IDENTIFICATION

1.8–3.0 mm. A small, round ladybug with a pale tan ground colour and a complex pattern of run-together small spots on the wing covers.

NOTES

This tiny ladybug is common in such microhabitats as the white, fungus-covered leaves of Saskatoons, maples, willows, and birches. It is important to note that sometimes mildew grows on aphid honeydew, and thus, it is common to find the wee-tiny ladybug in the company of aphid-eating species when using a sweep net or a beating sheet. In part because of its feeding habits, this species and its close relatives were traditionally placed in a separate tribe, the Psylloborini. However, recent molecular studies by Lori Shapiro and her colleagues have shown that the psylloborines are simply modified coccinellines with a taste for a different sort of food. However, I still place them at the end of the classification here, if only for ecological reasons.

APPENDIX A

Checklist

The following checklist enumerates the 75 species of ladybugs presently known from Alberta (plus the mealybug destroyer, which only lives in greenhouses). One of these, the uncus ladybug (indicated by an asterisk) was likely a one-time occurrences, represented by a single specimen that was probably transported here by accident.

To illustrate the progress we have made over the years, F.S. Carr's 1920 list included only 24 species (for northern Alberta only), including three that are no longer recognized as distinct. This was an improvement over Charles W. Leng's catalog of North American beetles, from the same year, in which one could find about 10 species of ladybugs that might inhabit Alberta, as part of the "Hudson's Bay Territory." In 1976, Joe Belicek listed 64 species from the province. In 1985, Bob Gordon also listed 64 species, although his list was different from Belicek's. In the *Checklist of Beetles of Canada and Alaska*, published in 1991, J. McNamara lists a whopping 82 species from Alberta, 11 of which I consider unsubstantiated, probably representing errors of identification. Is the checklist now stable? I doubt it, but the process of refining this list will be an enjoyable one for all of us with an interest in ladybugs. The lesser ladybugs in particular are sure to provide additional surprises.

Family Coccinellidae
    Subfamily Sticholotidinae
        Tribe Microweiseini
            Micro ladybug: *Microweisia misella* (LeConte)
    Subfamily Scymninae
        Tribe Stethorini
            American hairy ladybug: *Stethorus punctum* (LeConte)
            Newcomer hairy ladybug: *Stethorus punctillium* Weise
        Tribe Scymnini
            Mealybug destroyer: *Cryptolaemus montrouzieri* Mulsant
            Twice-stained ladybug: *Didion punctatum* (Melsheimer)
            Angular ladybug: *Didion longulum* Casey

Subgenus *Scymnus* (*Scymnus*)
    Apicanus ladybug: *Scymnus apicanus* Chapin
    Paracanus ladybug: *Scymnus paracanus* Chapin
    Opaque ladybug: *Scymnus opaculus* Horn
Subgenus *Scymnus* (*Pullus*)
    Fake Opaque ladybug: *Scymnus postpictus* Casey
    Carr's ladybug: *Scymnus carri* Gordon
    Uncus ladybug*: *Scymnus uncus* Wingo
    Diamond City ladybug: *Scymnus aquilonarius* Gordon
    Lacustrine ladybug: *Scymnus lacustris* LeConte
Ornate ladybug: *Nephus ornatus* (LeConte)
Farmer's ladybug: *Nephus georgei* (Weise)
Sordid ladybug: *Nephus sordidus* (Horn)

Tribe Selvadiini
  Tinytan ladybug: *Selvadius nunenmacheri* Gordon

Tribe Hyperaspidini
  Mimic ladybug: *Hyperaspidius mimus* Casey
  Well-marked ladybug: *Hyperaspidius insignis* Casey
  Vittate ladybug: *Hyperaspidius vittigerus* (LeConte)
  Hercules ladybug: *Hyperaspidius hercules* Belicek
  Unnamed ladybug: *Hyperaspidius* sp.
  Convivial ladybug: *Hyperaspis conviva* Casey
  Lugubrius ladybug: *Hyperaspis lugubris* (Randall)
  Lateral ladybug: *Hyperaspis lateralis* Mulsant
  Fastidious ladybug: *Hyperaspis fastidiosa* Casey
  Curved ladybug: *Hyperaspis inflexa* Casey
  Postica ladybug: *Hyperaspis postica* LeConte
  Oregon ladybug: *Hyperaspis oregona* Dobzhansky
  Blotch-backed ladybug: *Hyperaspis disconotata* Mulsant
  Undulate ladybug: *Hyperaspis undulata* (Say)
  Poorly-known ladybug: *Hyperaspis consimilis* LeConte
  Four-streaked ladybug: *Hyperaspis quadrivittata* LeConte
  Jasper ladybug: *Hyperaspis jasperensis* Belicek

Tribe Brachiacanthini
  Pale anthill ladybug: *Brachiacantha albifrons* (Say)
  Ursine anthill ladybug: *Brachiacantha ursina* (Fabricius)

Subfamily Chilocorinae
  Tribe Chilocorini
    Winter ladybug: *Brumoides septentrionis* (Weise)
    Round black ladybug: *Exochomus aethiops* (Bland)
    Twice-stabbed ladybug: *Chilocorus stigma* (Say)
    Once-squashed ladybug: *Chilocorus hexacyclus* Smith

Subfamily Coccidulinae
    Tribe Coccidulini
        Snow ladybug: *Coccidula lepida* LeConte
Subfamily Coccinellinae
    Tribe Coccinellini
        Marsh ladybug: *Anisosticta bitriangularis* (Say)
        Episcopalian ladybug: *Macronaemia episcopalis* (Kirby)
        Thirteen-spot ladybug: *Hippodamia tredecimpunctata* (Linnaeus)
        American ladybug: *Hippodamia americana* Crotch
        Waterside ladybug: *Hippodamia falcigera* Crotch
        Parenthesis ladybug: *Hippodamia parenthesis* (Say)
        Expurgate ladybug: *Hippodamia expurgata* Casey
        Five-spot ladybug: *Hippodamia quinquesignata* (Kirby)
        Glacial ladybug: *Hippodamia glacialis* (Fabricius)
        Sorrowful ladybug: *Hippodamia moesta* LeConte
        Convergent ladybug: *Hippodamia convergens* Guérin
        Colonel Casey's ladybug: *Hippodamia caseyi* Johnson
        Boulder ladybug: *Hippodamia oregonensis* Crotch
        Sinuate ladybug: *Hippodamia sinuata* Mulsant
        Flying saucer ladybug: *Anatis rathvoni* (LeConte)
        Large orange ladybug: *Anatis lecontei* Casey
        American eyespot ladybug: *Anatis mali* (Say)
        Streaked ladybug: *Myzia pullata* (Say)
        Subvittate ladybug: *Myzia subvittata* (Mulsant)
        Polkadot ladybug: *Calvia quatuordecimguttata* (Linnaeus)
        Two-spot ladybug: *Adalia bipunctata* (Linnaeus)
        Three-banded ladybug: *Coccinella trifasciata* Mulsant
        Transverse ladybug: *Coccinella transversoguttata* Faldeman
        Seven-spot ladybug: *Coccinella septempunctata* Linnaeus
        Nine-spot ladybug: *Coccinella novemnotata* Herbst
        High-country ladybug: *Coccinella alta* Brown
        Tamarack ladybug: *Coccinella monticola* Mulsant
        Hieroglyphic ladybug: *Coccinella hieroglyphica* Linnaeus
        Polished ladybug: *Cycloneda polita* Casey
        Halloween ladybug: *Harmonia axyridis* (Pallas)
        Painted ladybug: *Mulsantina picta* (Randall)
        Hudsonian ladybug: *Mulsantina hudsonica* (Casey)
        Wee-tiny ladybug: *Psyllobora vigintimaculata* (Say)

The author and his beating net, searching for lesser ladybugs in the valley of the Red Deer River near Jenner.

APPENDIX B

Glossary

abdomen: The hindmost of the three main body parts of an insect, usually hidden beneath the wing covers.
antenna (plural: antennae): A segmented feeler that extends from the forehead of an insect, separate from the mouthparts.
anterior: Toward the head end of the body.
antero-posteriorly: Running along the long axis of the body front-to-back.
apical (apex): On an appendage, the area at the tip and away from the body.
arboreal: Living in trees.
basal: On an appendage, the area close to where the structure joins the body.
biodiversity: The diversity of life on earth, including species numbers as well as general measures of genetic diversity and ecological diversity.
biologist: Someone who studies living things.
bugster: Informal term for insect enthusiast.
chalaza (plural: chalazae): A small bump on the outer surface of a ladybug larva, bearing a single seta.
chalazal base: The base of a chalaza, not including the seta.
coleopterist: A scientist who studies beetles.
cryptic species: Species that are difficult or impossible to recognize on the basis of external appearance.
cryptic: Hidden, or coloured in a fashion that enhances camouflage.
database: A computer-based collection of information on the occurrence of individual ladybugs in time and space.
diapause: A period of dormancy, which is necessary for an insect to complete its development to adulthood.
dispersal: A movement away from the place where a ladybug first emerges as an adult.
distal: On an appendage, the part farthest away from the body.
DNA analysis: Comparison of the sequence of the component parts of equivalent fragments of the genetic material (DNA—deoxyribose-nucleic acid) of ladybugs to estimate how closely different species are related.
dorsal: Referring to the top of the body, the "back."
dorsolateral: Positioned on the sides of the back, away from the midline.

eclose: To emerge from the pupa as an adult.
ecology: The study of how organisms interact with their surroundings, both living and nonliving.
elytra (singlular: elytron): The wing covers, formed from the hardened first pair of wings.
entomology: The study of insects.
evolutionary tree: A diagram showing how various evolutionary lineages are related. The tree shape comes from the fact that all lineages can trace their ancestry back to a common "trunk" in any given group of related organisms.
family: In classification, a grouping of genera.
faunistic: Of or relating to studies (like this book) that focus on the occurrence and ecology of the species in a particular group of related organisms that live in a particular area.
fecund: Capable of producing a great many young.
genus (plural: genera): In classification, a grouping of species.
geographic race: A subdivision of a species, in which the animals in a particular geographic area share recognizable differences from those elsewhere; the same as *subspecies.*
habitat: The sort of place (not the exact location) in which an organism lives.
instar: A period in the life of an insect (a larva in the case of a ladybug), between sheddings of the skin.
ladybugster: Informal term for a ladybug enthusiast.
larva: An immature insect, not yet a pupa or an adult.
lateral: Toward the sides of the body.
Latinized names: Names constructed in accordance with the rules of Latin grammar, even if they are derived from non-Latin roots.
marsh: A shallow, usually open and sunny wetland, largely vegetated by emergent plants with their roots under water and their stems and leaves above the surface.
maxillary palp (plural: palpae): the dorsal pair of "feelers" in the mouthparts of a ladybug or ladybug larvae.
mesonotum: The dorsal surface of the middle of the three segments that make up the thorax.
metanotum: The dorsal surface of the hindmost of the three segments that make up the thorax.
nocturnal: Active by night.
ocelli: On a ladybug larva, the eyes, which have only one lens each.
order: In classification, a grouping of families.
organism: A living thing.
parascolus (plural: parascoli): On a ladybug larva, a relatively large, but not branched, bump with many setae on it.
parasite: An organism that feeds on the tissues of another organism without necessarily killing the "host."

peatland: A wetland in which the decomposition of plants is slower than the accumulation of dead plants, such that peat is formed on the ground surface. The usual sorts of peatland in Alberta are called bogs and fens.

polymorphism: When the individuals in a single species can have two or more colour patterns or forms.

postcoxal line (= postcoxal arc): Thin lines on the underside of the first abdominal segment, which may be short (and thus "incomplete") or form a near semicircle (and are thus "complete"). The line can also be "deep" and extend almost to the second abdominal segment, or "shallow" and not extend that far to the rear.

posterior: Toward the hind or tail end of the body.

power (as in "10 power"): The amount of magnification provided by a lens, measured as the difference between the real and apparent length of an object; e.g., if the object appears ten times longer than it really is, then the lens is "10 power" or "10x."

prairie: The grassland region, which in Alberta extends south from about the latitude of Hannah and east from about the longitude of Calgary.

predatory: Habitually feeding on animals, which are killed and then eaten.

pronotal angles: The front and hind corners of the pronotum.

pronotal plate: The thickened, shield-like part of the pronotum of a ladybug larva, usually comprising most of the pronotum.

pronotum: The upper surface of the prothorax, which is the first of the three segments that make up the thorax.

prosternum: The lower surface of the prothorax, which is the first of the three segments that make up the thorax.

pupa: The resting stage of the life history of many insects, during which the larva transforms to an adult.

pupate: The process by which an insect larva becomes a pupa.

sagittal: Running along the imaginary midline of the body, between the right and left hand sides, around which the body is symmetrical.

segment: A component of a body part or of the body itself. Usually segments are lined up one behind the other and are separated from one another by a narrow ring of soft tissue, rather than the usual hardened exterior cuticle of an insect.

seta (plural: setae): Hair-like structures on the outside of an insect body that are set in sockets.

species: A group of organisms that not only look more or less alike, but are capable of interbreeding in nature and which form a more or less cohesive evolutionary lineage.

structural: Having to do with the anatomy of an organism.

subspecies: The same as *geographic race.*

systematics: The scientific study of how animals are related in an evolutionary sense and how this should be reflected in the way they are classified.

tarsal claws: The tiny claws on the tip of each leg. Adult ladybugs have two per leg, larvae have one.

taxonomy: Another term for *systematics*, without the same emphasis on evolutionary relationships.

tergum: The dorsal part of an insect body segment.

thorax: The second main body part of an insect, between the head and the abdomen, and comprising the prothorax and the pterothorax.

transverse: Running across the body, side to side, at right angles to the long axis.

ventral: Referring to the underside of the body, the belly or corresponding region of an appendage.

wing pads: On a pupa, the structures that are destined to become the wings.

APPENDIX C

Helpful Sources for Ladybug Study

BOOKS AND SUPPLIES

Bio Quip Inc.: a good selection of nets, other entomological equipment, books and videos.
address: 2321 Gladwick Street, Rancho Dominguez, CA, USA   90220
phone: (310) 667-8800   *fax*: (310) 667-8808   *email*: bqinfo@bioquip.com
website: http://www.bioquip.com/default.asp

Jean Paquet: insect nets and collection supplies.
address: 3 rue du Coteau, P.O. Box 953, Pont Rouge, QC   GH3 2E1
email: jeanpaquet@webnet.qc.ca
website: http://www.quebecinsectes.com/pages/pages_english/macrodontia_english.html

JOURNALS, WEBSITES, AND SOCIETIES

The Coleopterists Society: a group of entomologists (both professional and amateur) with a focus on beetles. The society meets informally once a year in conjunction with the Entomological Society of America and publishes The Coleopterists Bulletin four times a year. Membership is US $40 per year payable to Floyd Shockley, Department of Entomology, 413 Bio Sciences, University of Georgia, Athens, GA, USA   30602-2603.
website: http://www.coleopsoc.org/

Entomological Society of Alberta: a group of entomologists (primarily professional) that meets once a year for three days and publishes abstracts from papers presented at the meetings. Amateurs are welcome. Membership is $10 per year ($5 for students) payable to the Secretary of the Entomological Society of Alberta.
website: http://www.biology.ualberta.ca/courses.hp/esa/esa.htm

Entomological Society of Canada: a group of entomologists that meets once a year and publishes a newsletter as well as *The Canadian Entomologist*. Amateurs are welcome. Membership fee depends on whether you live in Canada or not, are a student or not, and which publications you wish to receive.
address: 93 Winston Avenue, Ottawa, ON K2A 1Y8
phone: (613) 725-2619 *fax*: (613) 725-9349 *email*: entsoc.can@bellnet.ca
website: http://www.esc-sec.org/

The E.H. Strickland Entomological Museum: houses a research collection of ladybugs and other insects, and has a great website.
address: Room 218, Earth and Atmospheric Sciences Building, University of Alberta, Edmonton, AB T6G 2E1
website: http://www.biology.ualberta.ca/facilities/strickland/

References

Alyokhin, A., and G. Sewell. 2004. Changes in a lady beetle community following the establishment of three alien species. Biol. Invasions, 6: 463–471.

Angalet, G.W., and R.L. Jacques. 1975. The establishment of *Coccinella septempunctata* L. in the continental United States. Coop. Econ. Ins. Rep. 25(45–48): 883–884.

Belicek, J. 1976. Coccinellidae of western Canada and Alaska with analyses of the transmontane zoogeographic relationships between the fauna of British Columbia and Alberta (Insecta: Coleoptera: Coccinellidae). Quaestiones Entomol. 12: 283–409.

Blatchley, W.S. 1910. An illustrated descriptive catalogue of the Coleoptera or beetles (exclusive of the Rhynchophora) known to occur in Indiana. Bull. Indiana Dep. Geol. Res. 1: 1–1386.

Block, V. 2001. A survey of the Coleoptera of the Canadian Shield region of northeastern Alberta. Report to Alberta Community Development Parks and Protected Areas Division, Edmonton. 13 pp.

Block, V. 2003. The Coleoptera of Colin–Cornwall Lakes Wildland Provincial Park. Report to Alberta Community Development Parks and Protected Areas Division, Edmonton. 11 pp.

Bousquet, Y. (ed.) 1991. Checklist of beetles of Canada and Alaska. Research Branch, Agriculture Canada, Ottawa, ON. Publ. No. 1861E. 430 pp.

Brown, W.J. 1962. A revision of the forms of *Coccinella* L. occurring in America north of Mexico (Coleoptera: Coccinellidae). Can. Entomol. 91: 785–808.

Cadotte, M.W., S.M. McMahon, and T. Fukami. (eds.). 2006. Conceptual ecology and invasion biology: reciprocal approaches to nature. Springer, Dordrecht, the Netherlands. 505 pp.

Dennett, D.C. 1991. Darwin's dangerous idea: evolution and the meanings of life. Simon & Schuster, New York. p. 63.

Dobzhansky, T. 1931. The North American beetles of the genus *Coccinella*. Proc. U.S. Natl. Mus. 80(34): 1–32.

Dobzhansky, T. 1967. The biology of ultimate concern. New American Library, New York.

Elliot, N., R. Kieckhefer, and W. Kauffman. 1996. Effects of an invading coccinellid on native coccinellids in an agricultural landscape. Oecologica, 105: 537–544.

Ellis, D.R., D.R. Prokrym, and R.G. Adams. 1999. Exotic lady beetle survey in northeastern United States: *Hippodamia variegata* and *Propylea quatuordecimpunctata* (Coleoptera: Coccinellidae). Entomol. News, 110: 73–84.

Elton, C.S. 1958. The ecology of invasions by animals and plants. Methuen, London. 181 pp.

Evans, E.W. 1991. Intra- *versus* intra-specific interactions of lady beetles (Coleoptera: Coccinellidae) attacking aphids. Oecologica, 87: 401–408.

Evans, E.W. 2000. Morphology of invasion: body size patterns associated with establishment of *Coccinella septempunctata* (Coleoptera: Coccinellidae) in western North America. Eur. J. Entomol. 97: 469–474.

Francis, F., E. Haubruge, P. Hastir, and C. Gaspar. 2001. Effect of the host plant on development and reproduction of the third trophic level, the predator *Adalia bipunctata* (Coleoptera: Coccinellidae). Environ. Entomol. 30: 947–952.

Frazer, B.D., and R.R. McGregor. 1992. Temperature-dependant survival and hatching rate of eggs of seven species of Coccinellidae. Can. Entomol. 124: 305–312.

Gage, J.H. 1920. The larvae of the Coccinellidae. Ill. Biol. Monogr. 62: 1–63.

Geoghegan, I.E., J.A. Chudek, R.L. MacKay, C. Lowe, S. Moritz, R.J. McNicol, A.N. Birch, G.H., and M.E.N. Majerus. 2000. Study of anatomical changes in *Coccinella septempunctata* (Coleoptera: Coccinellidae) induced by diet and by infection with the larva of *Dinocampus coccinellae* (Hymenoptera: Braconidae) using magnetic resonance microimaging. Eur. J. Entomol. 97: 457–461.

Gordon, R.D. 1976. The Scymnini (Coleoptera: Coccinellidae) of the United States and Canada: key to genera and revision of *Scymnus*, *Nephus*, and *Diomus*. Bull. Buffalo Soc. Nat. Sci. 28: 1–362.

Gordon, R.D. 1985. The Coccinellidae (Coleoptera) of America north of Mexico. J. N.Y. Entomol. Soc. 93: 1–912.

Gordon, R.D., and N. Vandenberg. 1995. Larval systematics of North American *Coccinella* L. (Coleoptera: Coccinellidae). Entomol. Scand. 26: 67–86.

Hamilton R.M., E.B. Dogan, G.B. Schaalje, and G.M. Booth. 1999. Olfactory response of the lady beetle *Hippodamia convergens* (Coleoptera: Coccinellidae) to prey related odors, including a scanning electron microscope study of the antennal sensilla. Environ. Entomol. 28: 812–822.

Harper, A.M., and C.E. Lilly. 1982. Aggregations and winter survival in southern Alberta of *Hippodamia quiquesignata* (Coleoptera: Coccinellidae), a predator of the pea aphid (Homoptera: Aphididae). Can. Entomol. 114: 303–309.

Harrison, W.C., and J. Acorn. 2000. The effects of the introduced lady beetle, *Coccinella septempunctata*, on the native coccinelline fauna of Alberta. Proc. Entomol. Soc. Alta. 48: 6.

Hesler, L.S., R.W. Kieckhefer, and D.A. Beck. 2001. First record of *Harmonia axyridis* (Coleoptera: Coccinellidae) in South Dakota and notes on its activity there and in Minnesota. Entomol. News, 112: 264–270.

Hodek, I., and A. Honěk (with contributions by P. Ceryngier and I. Kovar). 1996. Ecology of Coccinellidae. Series Entomologica, Vol. 54. Kluwer Academic Publishers, Dordrecht, the Netherlands. 464 pp.

Hodek, I., and P. Ceryngier. 2000. Sexual activity in Coccinellidae (Coleoptera): a review. Eur. J. Entomol. 97: 449–456.

Hoebeke, E.R., and A.G. Wheeler, Jr. 1980. New distribution records of *Coccinella septempunctata* L. in the eastern United States (Coleoptera: Coccinellidae). Coleopterists Bull. 34: 209–211.

Krafsur, E.S., P. Nariboli, and J.J. Obrycki. 1996. Gene flow and diversity at allozyme loci in the twospotted lady beetle (Coleoptera: Coccinellidae). Ann. Entomol. Soc. Am. 89: 410–419.

Kuznetsov, V. 1997. Lady beetles of the Russian Far East. Center For Systematic Entomology, Gainesville, FL. Memoir No. 1. 248 pp.

Lee, R.E. 1980. Aggregation of lady beetles on the shores of lakes (Coleoptera: Coccinellidae). Am. Midl. Nat. 104: 295–304.

Lucas, E., D. Coderre, and J. Brodeur. 2000. Selection of molting and pupation sites by *Coleomegilla maculata* (Coleoptera: Coccinellidae): avoidance of intraguild predation. Environ. Entomol. 29: 454–459.

Leng, C.W. 1920. Catalogue of the Coleoptera of America north of Mexico. John D. Sherman Jr., Mount Vernon, New York. 470 pp.

Majerus, M.E.N. 1994. Ladybirds. The new naturalist. HarperCollins, London. 367 pp.

Majerus, M.E.N., V. Strawson, and H. Roy. 2006. The potential impacts of the arrival of the harlequin ladybird, *Harmonia axyridia* (Pallas) (Coleoptera: Coccinellidae), in Britain. Ecol. Entomol. 31: 207–213.

Majerus, M.E.N., and P.W.E. Kearns. 1989. Ladybirds. Richmond Publishing, Slough, UK. Nat. Handb. Ser. No. 10. 103 pp.

McCorquodale, David B. 1998. Adventive lady beetles (Coleoptera: Coccinellidae) in eastern Nova Scotia, Canada. Entomol. News, 109: 15–20.

McMullen, R.D. 1967a. A field study of diapause in *Coccinella novemnotata* (Coleoptera: Coccinellidae). Can. Entomol. 99: 42–49.

McMullen, R.D. 1967b. The effects of photoperiod, temperature, and food supply on rate of development and diapause in *Coccinella novemnotata*. Can. Entomol. 99: 578–586.

Mondor, E.B., and J.L. Warren. 2000. Unconditioned and conditioned responses to colour in the predatory coccinellid, *Harmonia axyridis* (Coleoptera: Coccinellidae). Eur. J. Entomol. 97: 463–467.

Munyaneza, J., and J.J. Obrycki. 1998. Development of three populations of *Coleomegilla maculata* (Coleoptera: Coccinellidae) feeding on eggs of Colorado potato beetle (Coleoptera: Chrysomelidae). Environ. Entomol. 27: 117–122.

Nalepa, C., G.G. Kennedy, and C. Brownie. 2005. Role of visual contrast in the alighting behavior of *Harmonia axyridis* (Coleoptera: Coccinellidae) at overwintering sites. Environ. Entomol. 34: 425–431.

Obrycki, J.J., and T.J. Kring. 1998. Predaceous Coccinellidae in biological control. Annu. Rev. Entomol. 43: 295–321.

Obrycki, J.J., K.L. Giles, and A.M. Ormord. 1998a. Experimental assessment of interactions between larval *Coleomegilla maculata* and *Coccinella septempunctata* (Coleoptera: Coccinellidae) in field cages. Environ. Entomol. 27: 1280–1288.

Obrycki, J.J., K.L. Giles, and M. Ormond. 1998b. Interactions between an introduced and indigenous coccinellid species at different prey densities. Oecologia, 117: 279–285.

Pearson, D.L., C.B. Knisley, and C.J. Kazilek. 2006. A field guide to the tiger beetles of the United States and Canada. Oxford University Press, New York. 227 pp.

Phoofolo, M.W., and J.J. Obrycki. 1998. Potential for intraguild predation and competition among predatory Coccinellidae and Chrysopidae. Entomol. Exp. Appl. 89: 47–55.

Rees, B.E., D.M. Anderson, D. Bouk, and R.D. Gordon. 1994. Larval key to the genera and selected species of North American Coccinellidae Coleoptera. Proc. Entomol. Soc. Wash. 96: 387–412.

Riddick, Eric W., and Aldrich, J.R. 2004. Search for chemicals that modify the behavior of multicolored asian lady beetles. Am. Entomol. 50: 124–126.

Rodriguez-Saona, C., and J.C. Miller. 1999. Temperature-dependant effects on development, mortality, and growth of *Hippodamia convergens* (Coleoptera: Coccinellidae). Environ. Entomol. 28: 518–522.

Sax, D.F., J.J. Stchowicz, and S.D. Gaines (eds.). 2005. Species invasions: insights into ecology, evolution, and biogeography. Sinauer Associates, Inc. Publishers, Sunderland, Mass. 495 pp.

Schaefer, P.W., R.J. Dysart, and H.B. Specht. 1987. North American distribution of *Coccinella septempunctata* (Coleoptera: Coccinellidae) and its mass appearance in coastal Delaware. Environ. Entomol. 16: 368–373.

Sloggett, J.J., and M.E.N. Majerus. 2000. Habitat preference and diet in the predatory coccinellidae (Coleoptera): an evolutionary perspective. Biol. J. Linn. Soc. 70: 63–88.

Smith, G. 1959. The cytogenetic basis for speciation in Coleoptera. *In* Proceedings of the 10th International Congress of Genetics, Montreal (1956). pp. 444–450.

Snyder, W., S. Joseph, R.F. Preziosi, and A.Moore. 2000. Nutritional benefits of cannibalism for the lady beetle *Harmonia axyridis* (Coleoptera: Coccinellidae) when prey quality is poor. Environ. Entomol. 29: 1173–1179.

Tedders, W.L., and P.W. Schaefer. 1994. Release and establishment of *Harmonia axyridis* (Coleoptera: Coccinellidae) in the southeastern United States. Entomol. News, 105: 228–243.

Theodoropoulos, D.I. 1991. Invasion Biology: Critique of a Pseudoscience. Avvar Books, Blythe, Calif. 236 pp.

Turnock, B. 1996. Lady beetles on the Lake Manitoba beach. UFS (Delta Marsh) Annu. Rep. 31: 40–41.

Turnock, W.J., I.L. Wise, and F.O. Matheson. 2003. Abundance of some native coccinellines (Coleoptera: Coccinellidae) before and after the appearance of *Coccinella septempunctata*. Can. Entomol. 135: 391–404.

Vandenberg, N.J. 2002. Coccinellidae Latreille 1807. *In* R.H. Arnett Jr., M.C. Thomas, P.E. Skelley, and J.H. Frank (eds.). American beetles. Vol. 2. Polyphaga: Scarabaeoidea through Curculionoidea. CRC Press, Boca Raton, FL. pp. 371–389.

Van der Werf, W., E.W. Evans, and J. Powell. 2000. Measuring and modelling the dispersal of *Coccinella septempunctata* (Coleoptera: Coccinellidae) in alfalfa fields. Eur. J. Entomol. 97: 487–493.

Watson, W.Y. 1976. A review of the genus *Anatis* Mulsant (Coleoptera: Coccinellidae). Can. Entomol. 108: 935–944.

Wells, M.L., and R.M. McPherson. 1999. Population dynamics of three coccinellids in flue-cured tobacco and functional response of *Hippodamia convergens* (Coleoptera: Coccinellidae) feeding on tobacco aphids (Homoptera: Aphididae). Environ. Entomol. 28: 768–773.

Wheeler, A.G., and E.R. Hoebeke. 1995. *Coccinella novemnotata* in northeastern North America: historical occurrence and current status (Coleoptera: Coccinellidae). Proc. Entomol. Soc. Wash. 97: 701–716.

Wise, I.L., and W.J. Turnock. 2001. New records of coccinellid species for the province of Manitoba. Proc. Entomol. Soc. Man. 57: 5–10.

Xia, J.Y., W Van Der Werf, and R. Rabbinge. 1999. Temperature and prey density on bionomics of *Coccinella septempunctata* (Coleoptera: Coccinellidae) feeding on *Aphis gossypii* (Homoptera: Aphididae) on cotton. Environ. Entomol. 28: 307–314.